AF595104

Flavour Volatiles of Wine

Flavour Volatiles of Wine

Editors

Matteo Bordiga
Fulvio Mattivi

MDPI • Basel • Beijing • Wuhan • Barcelona • Belgrade • Manchester • Tokyo • Cluj • Tianjin

Editors

Matteo Bordiga
Università degli Studi del Piemonte
Orientale "A. Avogadro"
Italy

Fulvio Mattivi
University of Trento
Italy

Editorial Office

MDPI
St. Alban-Anlage 66
4052 Basel, Switzerland

This is a reprint of articles from the Special Issue published online in the open access journal *Foods* (ISSN 2304-8158) (available at: https://www.mdpi.com/journal/foods/special_issues/Flavour_Volatiles_Wine).

For citation purposes, cite each article independently as indicated on the article page online and as indicated below:

LastName, A.A.; LastName, B.B.; LastName, C.C. Article Title. *Journal Name* **Year**, *Volume Number*, Page Range.

ISBN 978-3-0365-2972-1 (Hbk)
ISBN 978-3-0365-2973-8 (PDF)

Contents

About the Editors

Matteo Bordiga is currently an Assistant Professor of Food Chemistry at Università del Piemonte Orientale (UPO), Novara, Italy. He earned his PhD in Food Science and MS in Chemistry and Pharmaceutical Technologies from the same University. His main research activity concerned food chemistry, investigating the different classes of polyphenols from an analytical, technological and nutritional point of view. More recently, he moved his research interests to wine chemistry, focusing his attention on the entire production process. He has published more than 50 research papers in peer-reviewed international and national journals. He is currently Editor of LWT—Food Science and Technology (IF 4.952; 2021), Elsevier. Since 2013, he has been an Editorial Board Member of International Journal of Food Science and Technology (IF 3.713; 2021), Wiley.

Fulvio Mattivi is currently a Full Professor of Food Chemistry at the University of Trento, Italy, where he is teaching in the Centre for Agriculture, Food and the Environment (C3A). Fulvio Mattivi's main research activities concern food chemistry, investigating different classes of polyphenols from the analytical, technological and nutritional point of view. More recently, his research interests have been directed toward plant biochemistry and mass-spectrometry-based metabolomics, and he is now also coordinating biochemical studies in the fields of plant, animal and nutritional metabolomics. He has published more than 200 research papers in peer-reviewed international journals. FM is the inventor of three Italian and two European patents and is among the founders of one academic spinoff. He is currently an Associated Editor of the scientific journals Oeno One and Molecules. He is on the Editorial Board of oenology journals in France, Germany and Austria, and nutritional publications (Frontiers in Pharmacology, Nutrition and Aging, and Recent Patents on Food, Nutrition and Agriculture).

 foods MDPI

Editorial

Special Issue on Flavour Volatiles of Wine

Matteo Bordiga [1,*] and Fulvio Mattivi [2,3]

1 Department of Pharmaceutical Sciences, Università degli Studi del Piemonte Orientale "A. Avogadro", Largo Donegani 2, 28100 Novara, Italy
2 Department of Cellular, Computational and Integrative Biology (CIBIO), University of Trento, Via Sommarive 9, Povo, 38123 Trento, Italy; fulvio.mattivi@unitn.it
3 Department of Food Quality and Nutrition, Research and Innovation Centre, Fondazione Edmund Mach, 38098 San Michele all'Adige, Italy
* Correspondence: matteo.bordiga@uniupo.it; Tel.: +39-032-137-5873

The perception of wine flavour and aroma is the result of several interactions between a large number of chemical compounds and sensory receptors. Compounds show synergistic (one compound enhances the perception of another) and antagonistic (one compound suppresses the perception of another) interactions. The chemical profile of a wine is derived from the entire process, starting from the grapes until bottled ageing. At the moment, wine makers are limited as to the range of yeasts that are able to impart some specific aromatic characteristic to a wine. Research focuses on issues such as adjusting the levels of flavour and aroma compounds, in particular esters and alcohols, producing enzymes that will release additional volatile compounds from the grapes, and reducing the amount of alcohol to levels that allow a better perception and release in the headspace of aroma and flavour compounds. New yeast strains are continuously being developed by traditional breeding techniques, leading to different flavour and aroma profiles in wine. In this context, the aim of the present Special Issue was to invite colleagues to submit their original research or review articles covering novel aspects of volatile compound research in the wine sector. Potential flavour volatiles of wine include: (i) varietal; (ii) pre-fermentative; (iii) formed by the yeast during alcoholic directly related to alcoholic fermentation; (iv) related to amino acid metabolism; (v) formed during malolactic fermentation; (vi) formed during ageing (reductive and oxidative pathway) and maturation. The aim was also to reach a mechanistic understanding of these pathways, with a focus on the reactions involved in the formation or degradation of key wine odorants and of the technological factors involved during the winemaking process.

Citation: Bordiga, M.; Mattivi, F. Special Issue on Flavour Volatiles of Wine. *Foods* **2022**, *11*, 69. https://doi.org/10.3390/foods11010069

Received: 11 November 2021
Accepted: 29 November 2021
Published: 29 December 2021

As a result, the final SI results in a balanced collection of original scientific research papers covering a broad range. A study reported the aroma profiles of withered Corvina and Corvinone wines from two different Valpolicella terroirs in relationship to yeast strain and the use of spontaneous fermentation [1]. Interestingly, in this study, the spontaneous fermentation reduced the sensory properties associated with the grape origin and variety, possibly due to the overproduction of acetic acid and ethyl acetate. Another study evaluated the effect of microwave treatment in grape maceration (at a laboratory scale) on the content of free and glycosidically bound varietal compounds of must and wines and on the overall aroma of wines produced with and without SO_2 [2]. In the study performed by Amores-Arrocha et al. (2021), a comparative study using bee pollen versus commercial fermentation activators in white and red winemaking evaluated promising new applications of this natural product [3].

Meanwhile, in order to differentiate white wines from Croatian indigenous varieties, the volatile aroma compounds were isolated by headspace solid-phase microextraction (HS-SPME) and analysed by comprehensive two-dimensional gas chromatography with time-of-flight mass spectrometry (GC×GC-TOF-MS) and conventional one-dimensional GC-MS [4]. More than 1000 compounds were detected and ca. 350 annotated. Several monoterpenes were proven particularly useful to differentiate among the wines.

Another study, for example, reports noteworthy data on the composition and sensory profiles of white wines made from the novel grape genotypes Albillo Dorado and Montonera del Casar (*Vitis vinifera* L.) [5]. Interestingly, Abreu et al. (2021) realised a comprehensive review reporting the description of the most important technological, chemical and sensory characteristics of some of the main fortified wines: Madeira, Port, Sherry, Muscat, and Vermouth [6].

This SI, therefore, provides a state-of-the-art overview of novel aspects of volatile compound research in the wine sector. We hope it will attract the interest not only among researchers and winemakers but also among the students in viticulture and oenology anxious to cover results in the frontiers of research that are not yet covered in the textbooks. This SI could not have been realised without the inputs of our valued contributors, all experts in the field, and the editorial team. We thank them for their willingness to contribute. Likewise, we are most indebted for the critical reviews from our peer reviewers.

Funding: This research received no external funding.

Conflicts of Interest: The authors declare no conflict of interest.

References

1. Luzzini, G.; Slaghenaufi, D.; Ugliano, M. Volatile Compounds in Monovarietal Wines of Two Amarone Della Valpolicella Terroirs: Chemical and Sensory Impact of Grape Variety and Origin, Yeast Strain and Spontaneous Fermentation. *Foods* **2021**, *10*, 2474. [CrossRef] [PubMed]
2. Muñoz García, R.; Simancas, R.O.; Díaz-Maroto, M.C.; Alañón Pardo, M.E.; Pérez-Coello, M.S. Effect of Microwave Maceration and SO2 Free Vinification on Volatile Composition of Red Wines. *Foods* **2021**, *10*, 1164. [CrossRef] [PubMed]
3. Amores-Arrocha, A.; Sancho-Galán, P.; Jiménez-Cantizano, A.; Palacios, V. A Comparative Study on Volatile Compounds and Sensory Profile of White and Red Wines Elaborated Using Bee Pollen versus Commercial Activators. *Foods* **2021**, *10*, 1082. [CrossRef]
4. Lukić, I.; Carlin, S.; Vrhovsek, U. Comprehensive 2D Gas Chromatography with TOF-MS Detection Confirms the Matchless Discriminatory Power of Monoterpenes and Provides In-Depth Volatile Profile Information for Highly Efficient White Wine Varietal Differentiation. *Foods* **2020**, *9*, 1787. [CrossRef] [PubMed]
5. Pérez-Navarro, J.; Izquierdo-Cañas, P.M.; Mena-Morales, A.; Chacón-Vozmediano, J.L.; Martínez-Gascueña, J.; García-Romero, E.; Hermosín-Gutiérrez, I.; Gómez-Alonso, S. Comprehensive Chemical and Sensory Assessment of Wines Made from White Grapes of Vitis vinifera Cultivars Albillo Dorado and Montonera del Casar: A Comparative Study with Airén. *Foods* **2020**, *9*, 1282. [CrossRef] [PubMed]
6. Abreu, T.; Perestrelo, R.; Bordiga, M.; Locatelli, M.; Coïsson, J.D.; Câmara, J.S. The Flavor Chemistry of Fortified Wines—A Comprehensive Approach. *Foods* **2021**, *10*, 1239. [CrossRef]

Article

Volatile Compounds in Monovarietal Wines of Two Amarone Della Valpolicella Terroirs: Chemical and Sensory Impact of Grape Variety and Origin, Yeast Strain and Spontaneous Fermentation

Giovanni Luzzini, Davide Slaghenaufi and Maurizio Ugliano *

Department of Biotechnology, University of Verona, Villa Lebrecht, via della Pieve 70, 37029 San Pietro in Cariano, Italy; giovanni.luzzini@univr.it (G.L.); davide.slaghenaufi@univr.it (D.S.)
* Correspondence: maurizio.ugliano@univr.it

Abstract: Aroma profiles of withered Corvina and Corvinone wines from two different Valpolicella terroirs were investigated in relationship to yeast strain and use of spontaneous fermentation. The results indicated that volatile chemical differences between wines were mainly driven by grape origin, which was associated with distinctive compositional profiles. Wine content in terpenes, norisoprenoids, benzenoids and C_6 alcohols, as well as some fermentative esters, were indeed significantly affected by grape origin. Conversely, yeast strain influence was mainly associated with fermentation-derived esters. Sensory analysis, besides confirming the major role of grape origin as driver of wine differentiation, indicated that spontaneous fermentations reduced the sensory differences associated with grape origin and variety, mainly due to high content of acetic acid and ethyl acetate.

Keywords: grape withering; yeast selection; terroir; spontaneous fermentation; Amarone della Valpolicella; ethyl acetate

Citation: Luzzini, G.; Slaghenaufi, D.; Ugliano, M. Volatile Compounds in Monovarietal Wines of Two Amarone Della Valpolicella Terroirs: Chemical and Sensory Impact of Grape Variety and Origin, Yeast Strain and Spontaneous Fermentation. *Foods* **2021**, *10*, 2474. https://doi.org/10.3390/foods10102474

Academic Editor: Beatriz Cancho

Received: 10 September 2021
Accepted: 13 October 2021
Published: 15 October 2021

1. Introduction

Valpolicella is an Italian wine region, well-known for the production of premium red wines. A peculiar feature of this region is the widespread use of post-harvest grape withering for the production of red wines, in particular the dry red passito Amarone della Valpolicella. Another characteristic of Valpolicella is the unique blend of grape varieties used for the production of its Protected Designation of Origin (PDO) wines, including the two main varieties Corvina and Corvinone, to be blended along with a range of other minor varieties [1].

From a geographical point of view, Valpolicella encompasses a pedo-climatically diverse territory, within which three different terroirs are also identified, namely the larger Valpolicella Classica (north-west of the city of Verona) and Valpolicella DOC (north-east of the city of Verona), and the smaller Valpantena (north of Verona) [1]. Several recent studies have provided novel insights into the chemical characteristics of Valpolicella wines and the contribution of withering and other technological factors to their composition [2–7]. Differences in grape composition due to variations in grape area of origin at both macro- and micro-scale have been shown to induce major changes in Valpolicella wine aroma composition [4,5]. Data concerning the relevance of terroir to wines for Amarone production are however scarce, in spite of the primary commercial relevance of this product.

Among wine constituents, volatile compounds play a central role in defining wine aroma and consequently its sensorial identity. Wine aroma is the product of a biochemical and technological sequence [8,9] resulting from the contribution of different volatile molecules deriving from grapes, fermentations, and reactions linked to aging, and sometimes oak and other woods. To date, more than 800 volatile compounds such as alcohols, esters, phenols, monoterpenes, norisoprenoids, lactones, aldehydes and ketones have been identified [10,11]. The majority of wine grapes are considered non-aromatic varieties in which many of the aroma metabolites that are key to wine aroma are present in various precursor forms, with grape variety, vineyard micro-climate and training deeply affecting their occurrence [12,13]. During fermentation, yeast activity combined with acid-catalyzed reactions release some of these compounds, while other fermentation-derived volatiles such as alcohols, fatty acids and esters are also produced [14]. As a whole, under equal fermentation conditions, the resulting wine aroma would arise from the interactions between precursors and nutrient levels of the grapes (in turn related to vineyard factors) and yeast enzymatic capabilities. Different studies indicate that the levels of compounds considered to be primarily of varietal origin, such as terpenes, can also be influenced by enzymatic activities of yeasts [14,15], whereas the levels of compounds considered to be of fermentation origin, such as esters, are influenced by grape composition [7,16,17]. Accordingly, there is a generalized interest in rationalizing the relative contribution of grape origin and composition as well as of yeast strain to the expression of wine aroma composition and olfactory characteristics. Nowadays, *Saccharomyces cerevisiae* starter cultures are largely used in winemaking to limit the growth of indigenous microorganisms and achieve more predictable and desired outcomes [5,18–24]. Nevertheless, a growing interest in non-*Saccharomyces* yeasts and spontaneous fermentation is currently observed, primarily with the aim of exploiting their metabolic diversity and obtaining even more diversified aroma profiles [25,26].

To date, most of the studies concerning the influence of grape terroir and/or yeast strain on red wine aroma composition have been carried out on wines from non-withered grapes. However, in the context of Amarone production, the traditional practice of post-harvest withering is inducing a weight loss of approximately 30%, with major consequences for grape and wine composition. First, during withering, grape metabolism is still active and a number of metabolic changes are observed beyond the simple concentration effect due to water evaporation, including increase or decrease in the content of certain aroma compounds and precursors [27,28]. The question arises, therefore, as to whether the recently reported observations concerning the existence, in Valpolicella wines from non-withered grapes, of aroma patterns associated with grape terroir of origin [5], are still relevant in the context of a passito red wine such as Amarone. Second, the water evaporation associated with withering results in the concentration of major grape components, primarily sugars. This has major implications for yeast behavior during fermentation, affecting different enzymatic activities associated with biosynthesis of volatile compounds and increasing the relevance of metabolic phenomena such as osmotic stress [6,29–31].

This research paper investigated the volatile and sensory characteristics of Corvina and Corvinone wines for Amarone production from the two main terroirs of Valpolicella, in relationship to different *S. cerevisiae* strains as well as to spontaneous fermentations. The main goal was to explore the relationship between grape composition/origin and fermentation management approaches, and to unravel their respective contribution to the expression of Amarone aroma chemical and olfactive profiles.

2. Materials and Methods

2.1. Grape Origins and Winemaking

Wines were produced with withered Corvina (*Vitis vinifera* L. cv. Corvina) or Corvinone (*Vitis vinifera*, L. cv. Corvinone) varieties. Grapes were harvested in September 2018 in vineyards belonging to the same winery and located in two terroirs within the Valpolicella appellation, namely Valpolicella DOC (Area 1) and Valpolicella Classica (Area 2). Grapes from Area 1 were obtained from three vineyard parcels located in the same estate near the town of Mezzane (45°30′36.7″ N, 11°08′02.7″ E). In the case of Area 2, two vineyard parcels were considered, located at a distance of approximately five kilometres from each other, in the towns of San Pietro in Cariano (45°30′40.5″ N, 10°54′58.6″ E) and San Giorgio in Valpolicella (45°32′16.6″ N, 10°51′19.4″ E), respectively. After harvesting, grapes were stored in a traditional warehouse (fruttaio) for withering. Sugar levels at harvest were in the range 195.2–207.7 g/L (Table S1). Withering lasted twelve weeks, with a gradual temperature decrease (from 16 °C to 7 °C) and a progressive increase in relative humidity (from 55% to 80%). When weight loss was approximately 30%, the grapes were pooled together in order to create two distinct vinification batches, including grapes from the parcels of Area 1 and Area 2 respectively. Grapes were manually destemmed and the berries randomized to obtain batches of 20 kg each. From each batch, eight hundred grams were taken, placed in a plastic bag and hand crushed with 80 mg of potassium metabisulphite and put into a 1.5 L glass vessel. Analytical parameters of the musts are provided in Table 1. Fermentations were carried out in duplicate with four different commercial yeasts, namely, *Saccharomyces cerevisiae* x *Saccharomyces kudriavzevi* AWRI 1503 (Yeast 1) (AB Mauri, Camellia, Australia), *Saccharomyces cerevisiae* AWRI 796 (Yeast 2) (AB Mauri, Camellia, Australia), *Saccharomyces cerevisiae* Zymaflore® XPURE (Yeast 3) (Laffort, Floirac, France), and *Saccharomyces cerevisiae* Zinfandel (Yeast 4) (Vason, Verona, Italy). Active dry yeast of each commercial starter was rehydrated in water at 37 °C for 15 min, then 1.6 mL of each culture (100 g/L) was used to inoculate individual grape batches. A fifth experimental modality was also prepared, consisting of a spontaneous fermentation without the addition of potassium metabisulphite (Spontaneous). All fermentations were carried out at 22 ± 1 °C, with cap being broken twice a day by gently pressing down skins with a steel plunger, and density, weight and temperature monitored daily. Upon completion of alcoholic fermentation (glucose-fructose < 2 g/L), wines were pressed with a ten litre stainless steel basket press and supplemented with potassium metabisulphite until a final free SO_2 concentration of 25 mg/L was achieved. Wines were then clarified by centrifugation at 4500 rpm for 15 min at 5° C (Avanti J-25, Beckman Coulter, CA, USA) and bottled in 330 mL glass bottles with crown caps, with free SO_2 concentration of 25 mg/L.

Table 1. Enological parameters of musts at crush.

	Glucose and Fructose (g/L)	pH	PAN [1] (mg/L)	AMMONIA (mg/L)	YAN [2] (mg/L)
	Mean ± sd	Mean ± sd	Mean ± sd	Mean ± sd	Mean ± sd
Area 1 Corvina	243.8 ± 3.1c	3.17 ± 0.03c	111.9 ± 7.8c	36.8 ± 2.4c	142.2 ± 9.3c
Area 2 Corvina	291.2 ± 3.6a	3.36 ± 0.04a	105.0 ± 5.4c	46.3 ± 6.2b	143.1 ± 6.4c
Area 1 Corvinone	235.1 ± 2.4d	3.02 ± 0.01d	149.3 ± 8.4a	73.9 ± 4.4a	210.1 ± 11.3a
Area 2 Corvinone	254.9 ± 4.2b	3.25 ± 0.01b	124.1 ± 9.1b	49.9 ± 3.6b	165.1 ± 11.8b

[1] PAN: primary amino nitrogen; [2] YAN: yeast assimilable nitrogen. Different letters in the same column denote statistically significant difference as obtained by Kruskal–Wallis (α = 0.05) with Dunn multiple pairwise comparison.

2.2. Main Enological Parameters

Glucose-fructose, ammonia, primary amino nitrogen (PAN), acetic acid, and total acidity (expressed in grams of tartaric acid) were analyzed using a Biosystems Y15 multiparametric analyzer (Sinatech, Fermo, Italy). YAN (yeast assimilable nitrogen) was obtained as the sum of PAN and ammonia. For each parameter, a specific kit (Sinatech, Fermo, Italy) was used. Ethanol was analyzed with an Alcolyzer dma 4500 (Anton Paar, Graz, Austria).

2.3. Analysis of Volatile Compounds

For quantification of alcohols, esters, fatty acids, and benzenoids (except methyl salicylate), SPE extraction followed by GC-MS analysis was used, following the procedure described by Slaghenaufi et al. [4]. An amount of 100 µL of internal standard 2-octanol (4.2 mg/L in ethanol) was added to samples prepared with 50 mL of wine and diluted with 50 mL of deionized water. Samples were loaded onto a BOND ELUT-ENV, SPE cartridge (Agilent Technologies. Santa Clara, CA, USA) previously activated with 20 mL of dichloromethane, 20 mL of methanol and equilibrated with 20 mL of water. After sample loading, the cartridges were washed with 15 mL of water. Free volatile compounds were eluted with 10 mL of dichloromethane, and then concentrated under gentle nitrogen stream to 200 µL prior to GC injection.

For quantification of terpenes, norisoprenoids, lactones and methyl salicylate, SPME extraction followed by GC-MS analysis was used, following the procedure described by Slaghenaufi et al. (2018) [32]. An amount of 5 µL of internal standard 2-octanol (4.2 mg/L in Ethanol) was added to 5 mL of wine diluted with 5 mL of deionized water in a 20 mL glass vial. An amount of 3 g of NaCl was added prior to GC-MS analysis. Samples were equilibrated for 1 min at 40 °C. Subsequently SPME extraction was performed using a 50/30 µm divinylbenzene–carboxen–polydimethylsiloxane (DVB/CAR/PDMS) fiber (Supelco, Bellafonte, PA, USA) exposed to sample headspace for 60 min. GC-MS analysis was carried out on an HP 7890A (Agilent Technologies) gas chromatograph coupled to a 5977B quadrupole mass spectrometer, equipped with a Gerstel MPS3 auto sampler (Müllheim/Ruhr, Germany). Separation was performed using a DB-WAX UI capillary column (30 m × 0.25, 0.25 µm film thickness, Agilent Technologies) and helium (6.0 grade) as carrier gas at 1.2 mL/min of constant flow rate. GC oven was programmed as follows: started at 40 °C for 3 min, raised to 230 °C at 4 °C/min and maintained for 20 min. Mass spectrometer was operated in electron ionization (EI) at 70 eV with ion source temperature at 250 °C and quadrupole temperature at 150 °C. Mass spectra were acquired in synchronous Scan (*m/z* 40–200) and SIM mode. Samples were analyzed in random order.

Calibration curves were prepared for both quantification methods. For SPE-GC-MS method, a calibration curve was prepared for each analyte using seven concentration points and three replicate solutions per point in model wine (12% *v*/*v* ethanol, 3.5 g/L tartaric acid, pH 3.5) 100 µL of internal standard 2-octanol (4.2 mg/L in ethanol) was added to each calibration solution, which was then submitted to SPE extraction and GC-MS analysis as described for the samples. For SPME-GC-MS method a calibration curve was prepared for each analyte using seven concentration points and three replicate solutions per point in red wines. An amount of 5 µL of internal standards 2-octanol (4.2 mg/L in ethanol) was added to each calibration solution, which was then submitted to SPME extraction and GC-MS analysis as described for the samples.

Calibration curves were obtained using Chemstation software (Agilent Technologies, Inc.) by linear regression, plotting the response ratio (analyte peak area divided by internal standard peak area) against concentration ratio (added analyte concentration divided by internal standard concentration). Retention indices, quantitation and qualifying ions are reported in Table S2.

2.4. Sensory Evaluation

Sensory evaluation of the experimental wines was carried out by means of the sorting task methodology, as described by Alegre et al. (2017) [33] with slight differences. Twelve judges (6 men and 6 women), wine science researchers or teaching staff regularly involved in winemaking and/or wine evaluation participated to the sessions. They were all considered wine experts according to Parr et al. (2002) [34] specifications. One hour before the test, samples were removed from the 16 °C cold room and 20 mL was poured in ISO wine glasses (https://www.iso.org/standard/9002.html, accessed on 1 September 2021) labelled with 3-digit random codes and covered by plastic Petri dishes; all samples were served at 22 ± 1 °C, and glasses were randomized for each panelist. Panelists were asked to sort the wines into groups based on odor similarities exclusively by orthonasal evaluation, with no request to indicate specific odor descriptors. They could make as many groups as they wished.

This study contains sensory analyses carried out by a trained panel which, based on local policy, does not require the specific approval of an ethics committee.

2.5. Statistical Analyses

Principal Component Analysis (PCA), Kruskal–Wallis (α = 0.05), and Hierarchical Cluster Analysis (HCA) were performed using XLSTAT 2017 (Addinsoft SARL, Paris, France). Heat map was performed with MetaboAnalyst v. 5.0 (http://www.metaboanalyst.ca, accessed on 16 March 2021), created at the University of Alberta, Edmonton, AB, Canada.

3. Results and Discussion

3.1. Main Enological Parameters

Enological parameters of withered grapes musts at crush are shown in Table 1. For each variety, Area 2 samples showed higher glucose and fructose than Area 1. pH varied across grape batches, ranging between 3.02 in Area 1 Corvinone, and 3.25 in Area 2 Corvinone. YAN content in Corvina was very similar between the two areas, while in Corvinone it was higher for Area 1.

These data reflected the natural variation in grape composition but also the effect of withering. The influence of the latter was strongly relevant in determining major differences between Areas 1 and 2, in particular for grape sugar content. While at harvest grape sugar content varied overall of less than 10 g/L (Table S1), after withering grapes of Area 2 generally exceeded those of Area 1 up to nearly 50 g/L of sugars. According to Barbanti et al. [35], withering kinetics can vary considerably depending on grape variety, bunch structure, as well as berry surface/volume ratio. In agreement with their observations, the net increase in sugar content observed here was lower in Corvinone, which is reported to have lower surface/volume ratio. Fermentation kinetics (Figure S1) showed small differences between commercial yeast strains, whereas spontaneous fermentations showed generally slower start and longer time to achieve dryness. Enological parameters of wines and their Kruskal–Wallis analysis are shown in Table 2, pH and total acidity were primarily affected by grape origins. Acetic acid was mainly influenced by yeast in both the varieties. Spontaneous fermentation showed much higher content of acetic acid than the other treatments, typically above 0.8 g/L. With an odor threshold of 0.7 g/L [36], a sensory involvement of acetic acid in several of the wines would be expected, which will be addressed later. Area 2 wines, in both varieties, in agreement with different glucose and fructose content of musts showed higher ethanol levels.

Table 2. Enological parameters of wines at the end of alcoholic fermentation (glucose and fructose <2 g/L).

		Yeast 1	Yeast 2	Yeast 3	Yeast 4	Spontaneous
		mean ± sd	mean ± sd	mean ± sd	mean ± sd	mean ± sd
Area 1 Corvina	Total acidity (g/L of tartaric acid)	6.7 ± 0.1b	7.7 ± 0.1a	6.7 ± 0.1b	7.4 ± 0.1a	7.8 ± 1.9a
	pH	3.01 ± 0.01b	3.08 ± 0.01a	3.07 ± 0.01a	3.07 ± 0.01a	3.09 ± 0.02a
	Acetic acid (g/L)	0.28 ± 0.02c	0.30 ± 0.00c	0.37 ± 0.04b	0.39 ± 0.01b	0.89 ± 0.01a
	Ethanol (% *v*/*v*)	14.77 ± 0.31a	14.69 ± 0.09a	14.82 ± 0.42a	14.66 ± 0.31a	14.65 ± 0.30a
Area 2 Corvina	Total acidity (g/L of tartaric acid)	7.6 ± 0.7b	10.3 ± 0.3a	7.8 ± 0.1b	7.6 ± 0.2b	7.8 ± 0b
	pH	2.98 ± 0.01a	2.99 ± 0.01a	2.98 ± 0.02a	2.98 ± 0.00a	2.92 ± 0.01b
	Acetic acid (g/L)	0.20 ± 0.11c	0.47 ± 0.18b	0.37 ± 0.18b	0.13 ± 0.01c	0.89 ± 0.06a
	Ethanol (% *v*/*v*)	17.91 ± 0.40a	17.73 ± 0.09a	17.86 ± 0.11a	17.68 ± 0.08a	17.41 ± 0.24a
Area 1 Corvinone	Total acidity (g/L of tartaric acid)	5.8 ± 0.1c	7.2 ± 0.2a	6.4 ± 0b	6 ± 0b	6.2 ± 0.3b
	pH	3.35 ± 0.06a	3.30 ± 0.01a	3.32 ± 0.01a	3.31 ± 0.02a	3.23 ± 0.01b
	Acetic acid (g/L)	0.50 ± 0.04c	0.60 ± 0.07b	0.59 ± 0.01b	0.44 ± 0.03d	0.85 ± 0.01a
	Ethanol (% *v*/*v*)	14.34 ± 0.17a	14.17 ± 0.35a	14.24 ± 0.31a	14.54 ± 0.09a	14.16 ± 0.02a
Area 2 Corvinone	Total acidity (g/L of tartaric acid)	6.6 ± 0.37b	7.81 ± 0.1a	6.62 ± 0.1b	7.04 ± 0.0b	6.84 ± 0.4b
	pH	3.21 ± 0.01a	3.22 ± 0.01a	3.22 ± 0.01a	3.23 ± 0.01a	3.21 ± 0.0a
	Acetic acid (g/L)	0.23 ± 0.01c	0.41 ± 0.21b	0.28 ± 0.02b	0.21 ± 0.02c	0.82 ± 0.01a
	Ethanol (% *v*/*v*)	15.50 ± 0.12a	15.41 ± 0.03a	15.31 ± 0.41a	15.32 ± 0.21a	15.51 ± 0.5a

Different letters in the same row denote statistically significant difference as obtained by Kruskal–Wallis (α = 0.05) with Dunn multiple pairwise comparison.

3.2. Volatile Compounds

The concentrations of all quantified volatile compounds, including esters, alcohols, fatty acids, terpenoids, norisoprenoids, and benzenoids, are reported in Tables 3–6. In Corvina wines (Tables 3 and 4), twenty-eight volatile compounds were found to be significantly different according to grape origin (α = 0.05) (Table S3). Of these, most were grape-derived compounds, such terpenes, norisoprenoids, benzenoids and C_6 alcohols, but some fermentative compounds, were also included. Nineteen volatile compounds, mainly alcohols, esters, acids showed statistically significant differences (α = 0.05) due to yeast strain/inoculation strategy. In the case of Corvinone wines (Tables 5 and 6), thirty-one compounds were significantly different according to grape area of origin ($\alpha = 0.05$) (Table S3). In addition, many ethyl esters were impacted by grape origin. Nineteen compounds showed significant differences ($\alpha = 0.05$) according to employed yeast: mainly alcohols, acids and esters.

Table 3. Concentration (μg/L) of volatile compounds of Area 1 Corvina wines.

	Yeast 1	Yeast 2	Yeast 3	Yeast 4	Spontaneous
	mean ± sd	mean ± sd	mean ± sd	mean ± sd	mean ± sd
Alcohols					
1-Butanol	371.05 ± 14.63	241.87 ± 9.91	186.51 ± 6.10	241.95 ± 3.30	146.94 ± 2.74
2-Butanol	4090.42 ± 127.27	4366.08 ± 114.18	7199.53 ± 88.06	4483.68 ± 86.56	4814.56 ± 106.43
1-Pentanol	41.94 ± 2.13	44.76 ± 0.83	46.75 ± 2.60	46.35 ± 3.87	52.11 ± 0.41
Isoamyl alcohol (mg/L)	260.1 ± 5.16	270.95 ± 10.67	294.79 ± 1.93	256.68 ± 4.26	123.5 ± 0.57
Phenylethyl alcohol (mg/L)	24.75 ± 0.92	17.69 ± 0.4	20.82 ± 0.61	23.56 ± 2.11	9.50 ± 0.15
Methionol	105.27 ± 4.96	84.92 ± 5.44	123.47 ± 14.62	143.71 ± 3.83	33.01 ± 1.29
C_6 alcohols					
1-Hexanol	954.18 ± 9.50	948.77 ± 8.58	902.40 ± 8.26	853.42 ± 15.13	768.00 ± 24.98
trans-3-Hexen-1-ol	9.30 ± 0.02	9.92 ± 0.94	10.55 ± 0.59	11.07 ± 0.26	6.79 ± 0.35
cis-3-Hexen-1-ol	43.27 ± 1.94	40.99 ± 0.93	44.51 ± 1.95	39.98 ± 1.09	40.64 ± 1.39
cis-2-Hexen-1-ol	18.26 ± 0.17	15.85 ± 0.70	16.41 ± 1.18	16.84 ± 0.79	16.23 ± 1.92
Acetate esters					
Isoamyl acetate	287.14 ± 4.84	417.00 ± 6.83	365.18 ± 25.42	342.91 ± 3.76	663.35 ± 12.79
n-Hexyl acetate	19.11 ± 0.08	51.72 ± 2.94	18.96 ± 0.12	17.36 ± 0.36	23.57 ± 1.29
2-Phenylethyl acetate	23.91 ± 2.09	18.75 ± 0.72	19.87 ± 0.37	23.83 ± 2.04	44.81 ± 0.71
Ethyl acetate (mg/L)	58.49 ± 2.7	83.66 ± 5.86	64.52 ± 7.29	41.86 ± 1.93	128.73 ± 3.26
Branched-chain fatty acids ethyl esters					
Ethyl 2-methylbutanoate	3.41 ± 0.12	2.73 ± 1.39	3.38 ± 0.01	3.17 ± 0.23	0.90 ± 0.14
Ethyl 3-methylbutanoate	2.57 ± 0.01	2.95 ± 0.19	2.73 ± 0.08	2.79 ± 0.21	1.05 ± 0.07
Fatty acids ethyl esters					
Ethyl butanoate	260.10 ± 29.42	422.74 ± 22.32	251.05 ± 11.19	184.12 ± 0.15	95.45 ± 1.34
Ethyl hexanoate	361.31 ± 15.52	555.12 ± 23.57	414.17 ± 24.71	271.97 ± 11.41	176.53 ± 5.61
Ethyl octanoate	171.19 ± 15.40	309.87 ± 24.84	247.56 ± 2.04	148.21 ± 7.71	89.70 ± 3.05
Ethyl lactate	225.35 ± 7.00	236.10 ± 20.09	284.54 ± 2.35	203.32 ± 12.91	359.76 ± 0.62
Ethyl decanoate	65.28 ± 5.73	42.70 ± 1.45	45.31 ± 2.35	14.25 ± 1.29	17.94 ± 0.08
Other esters					
Ethyl 3-hydroxybutanoate	148.56 ± 7.57	247.18 ± 15.42	227.89 ± 12.49	139.87 ± 6.30	78.60 ± 2.91
Ethyl 2-hydroxyhexanoate	0.42 ± 0.02	1.23 ± 0.16	1.52 ± 0.10	1.01 ± 0.00	0.53 ± 0.01
Fatty acids					
3-Methylbutanoic acid	424.86 ± 19.17	310.26 ± 9.60	439.13 ± 20.92	414.41 ± 11.50	547.27 ± 5.76
Hexanoic acid	1824.83 ± 50.37	2973.50 ± 117.13	2209.76 ± 14.81	1509.15 ± 23.80	941.09 ± 29.54
Octanoic acid	3537.02 ± 193.32	4631.18 ± 308.33	4100.89 ± 2.94	3219.52 ± 27.57	2216.20 ± 164.18
Terpenoids					
cis-Linaloloxide	0.55 ± 0.05	0.10 ± 0.01	0.08 ± 0.02	0.02 ± 0.00	0.68 ± 0.03
trans-Linaloloxide	0.96 ± 0.07	0.03 ± 0.01	0.05 ± 0.03	0.16 ± 0.13	0.62 ± 0.04
Linalool	5.76 ± 1.00	6.98 ± 0.42	5.81 ± 0.36	5.07 ± 0.13	5.06 ± 0.10
Geraniol	2.23 ± 0.13	3.50 ± 0.69	3.35 ± 0.62	2.67 ± 0.14	3.01 ± 0.32
α-Terpineol	3.50 ± 0.00	3.90 ± 0.52	2.82 ± 0.33	3.22 ± 0.13	2.81 ± 0.40
β-Citronellol	26.76 ± 3.07	8.57 ± 1.70	13.30 ± 1.49	16.31 ± 0.43	15.69 ± 1.77
α-Phellandrene	3.35 ± 0.27	3.66 ± 0.16	3.16 ± 0.06	1.26 ± 0.06	3.20 ± 0.14
α-Terpinen	0.13 ± 0.01	0.16 ± 0.01	0.11 ± 0.01	0.45 ± 0.07	0.14 ± 0.00
β-Myrcene	4.36 ± 0.35	5.65 ± 0.35	3.89 ± 0.23	2.83 ± 0.04	4.80 ± 1.10
Limonene	0.74 ± 0.04	0.93 ± 0.08	0.63 ± 0.04	0.67 ± 0.02	0.78 ± 0.16
1,4-Cineole	nd	nd	0.04 ± 0.00	0.43 ± 0.03	0.09 ± 0.01
1,8-Cineole	0.08 ± 0.00	0.04 ± 0.06	0.13 ± 0.01	0.05 ± 0.01	0.08 ± 0.01
p-Cymene	0.32 ± 0.01	0.39 ± 0.01	0.28 ± 0.02	0.52 ± 0.02	0.39 ± 0.01
Terpinolene	0.47 ± 0.08	0.59 ± 0.02	0.45 ± 0.07	0.52 ± 0.01	0.52 ± 0.04
Terpinen-4-ol	0.48 ± 0.04	0.47 ± 0.09	1.21 ± 0.15	14.11 ± 0.01	0.75 ± 0.07

Table 3. *Cont.*

	Yeast 1	Yeast 2	Yeast 3	Yeast 4	Spontaneous
Norisoprenoids					
β-Damascenone	8.33 ± 1.56	8.38 ± 1.17	5.09 ± 0.30	6.87 ± 0.16	6.01 ± 0.81
3-Hydroxy-β-damascone	0.21 ± 0.01	0.40 ± 0.04	0.28 ± 0.09	0.36 ± 0.04	0.30 ± 0.01
Vitispirane	3.45 ± 0.21	3.88 ± 0.18	2.66 ± 0.04	3.62 ± 0.02	3.10 ± 0.99
TPB	0.04 ± 0.01	0.04 ± 0.00	0.03 ± 0.01	0.04 ± 0.00	0.02 ± 0.00
TDN	0.73 ± 0.03	0.78 ± 0.13	0.50 ± 0.06	0.69 ± 0.03	0.57 ± 0.07
Benzenoids and others					
Benzyl alcohol	279.52 ± 13.31	296.78 ± 13.09	316.75 ± 7.83	307.61 ± 26.40	251.22 ± 11.33
Vanillin	5.50 ± 0.00	5.52 ± 0.12	5.43 ± 0.15	5.54 ± 0.16	2.00 ± 0.07
Ethyl vanillate	127.88 ± 5.50	119.02 ± 3.81	123.33 ± 3.86	124.45 ± 2.86	126.81 ± 5.37
Methyl vanillate	5.46 ± 0.20	8.72 ± 0.63	5.17 ± 0.76	6.23 ± 0.23	5.49 ± 0.52
Benzaldehyde	16.14 ± 1.19	20.31 ± 1.63	16.66 ± 0.59	17.85 ± 0.70	70.95 ± 7.71
Eugenol	7.01 ± 0.04	7.32 ± 0.02	7.26 ± 0.00	7.14 ± 0.22	6.45 ± 0.49
Methyl salicylate	0.99 ± 0.16	1.12 ± 0.31	0.63 ± 0.01	0.74 ± 0.05	0.39 ± 0.07
2,6-Dimethoxyphenol	5.30 ± 0.55	5.42 ± 0.37	6.29 ± 0.16	6.39 ± 0.18	5.02 ± 0.16
Furfural	1.38 ± 0.06	1.24 ± 0.09	1.89 ± 0.00	1.17 ± 0.18	0.69 ± 0.04
γ-Decalactone	2.36 ± 0.16	2.41 ± 0.13	2.27 ± 0.00	2.73 ± 0.01	2.15 ± 0.19
δ-Decalactone	32.99 ± 2.51	31.19 ± 1.33	28.78 ± 1.27	30.24 ± 3.59	23.71 ± 2.13

Nd means not detected. TPB is short for (E)-1-(2,3,6-Trimethylphenyl)-buta-1,3-diene. TDN is short for 1,6,-trimethyl-1,2-dihydronapthalene.

Table 4. Concentration (μg/L) and standard deviation (± μg/L) of volatile compounds of Area 2 Corvina wines.

	Yeast 1	Yeast 2	Yeast 3	Yeast 4	Spontaneous
	mean ± sd	mean ± sd	mean ± sd	mean ± sd	mean ± sd
Alcohols					
1-Butanol	441.60 ± 12.31	220.49 ± 9.52	190.27 ± 8.34	205.29 ± 9.91	290.30 ± 14.21
2-Butanol	3487.83 ± 19.06	4174.76 ± 194.86	4787.57 ± 1.14	3691.44 ± 374.75	5230.78 ± 139.41
1-Pentanol	36.54 ± 3.17	46.36 ± 2.33	47.77 ± 6.75	43.88 ± 3.17	53.30 ± 16.89
Isoamyl alcohol (mg/L)	224.67 ± 6.78	229.71 ± 18.30	223.20 ± 16.50	207.07 ± 31.67	230.41 ± 52.37
Phenylethyl alcohol (mg/L)	23.78 ± 0.78	21.43 ± 0.24	21.91 ± 0.45	21.16 ± 0.71	22.77 ± 0.71
Methionol	102.96 ± 3.90	98.07 ± 9.43	127.90 ± 5.51	153.40 ± 17.46	162.44 ± 107.20
C_6 alcohols					
1-Hexanol	838.43 ± 25.46	781.33 ± 10.08	712.71 ± 4.66	798.18 ± 16.11	763.07 ± 26.74
trans-3-Hexen-1-ol	7.30 ± 0.41	7.90 ± 0.24	6.53 ± 0.43	7.31 ± 0.42	6.82 ± 1.25
cis-3-Hexen-1-ol	38.18 ± 3.79	39.27 ± 1.43	34.32 ± 0.60	38.98 ± 1.32	39.13 ± 4.83
cis-2-Hexen-1-ol	16.33 ± 1.27	16.70 ± 0.50	15.82 ± 0.24	16.70 ± 0.08	16.24 ± 0.14
Acetate esters					
Isoamyl acetate	201.22 ± 3.98	310.27 ± 1.60	305.31 ± 3.94	302.26 ± 18.75	342.06 ± 165.76
n-Hexyl acetate	20.18 ± 1.35	38.97 ± 1.15	22.50 ± 2.48	22.36 ± 1.57	32.38 ± 3.58
2-Phenylethyl acetate	22.27 ± 0.41	25.38 ± 1.02	22.87 ± 1.31	14.91 ± 2.82	19.87 ± 0.33
Ethyl acetate (mg/L)	51.97 ± 5.39	62.89 ± 7.08	52.04 ± 0.48	48.51 ± 1.12	153.00 ± 14.63
Branched-chain fatty acids ethyl esters					
Ethyl 2-methylbutanoate	3.50 ± 0.09	3.86 ± 0.68	3.99 ± 0.11	2.22 ± 0.13	1.80 ± 0.33
Ethyl 3-methylbutanoate	4.07 ± 0.08	3.96 ± 0.09	4.08 ± 0.21	3.83 ± 0.10	3.97 ± 0.08
Fatty acids ethyl esters					
Ethyl butanoate	232.99 ± 3.66	244.30 ± 3.06	170.31 ± 6.74	160.74 ± 7.11	172.36 ± 21.39
Ethyl hexanoate	342.78 ± 2.55	391.36 ± 10.20	280.57 ± 14.14	232.48 ± 3.34	265.86 ± 95.90
Ethyl octanoate	232.65 ± 16.50	296.96 ± 3.34	201.46 ± 6.02	152.80 ± 1.60	182.44 ± 49.89
Ethyl lactate	244.39 ± 18.96	246.63 ± 7.59	236.89 ± 3.22	193.08 ± 7.44	336.42 ± 28.96
Ethyl decanoate	70.89 ± 1.82	82.04 ± 7.63	56.46 ± 3.37	29.78 ± 6.25	36.49 ± 4.51
Other esters					
Ethyl 3-hydroxybutanoate	105.64 ± 6.74	134.61 ± 0.52	65.35 ± 0.66	77.05 ± 16.40	83.53 ± 2.35
Ethyl 2-hydroxyhexanoate	0.25 ± 0.02	0.54 ± 0.06	0.43 ± 0.00	0.46 ± 0.04	0.36 ± 0.24
Fatty acids					
3-Methylbutanoic acid	318.40 ± 7.15	328.26 ± 33.71	388.00 ± 3.51	319.32 ± 21.51	269.95 ± 123.81
Hexanoic acid	2030.04 ± 44.08	1844.35 ± 46.25	1451.38 ± 138.40	1316.03 ± 152.40	1338.87 ± 293.02
Octanoic acid	4261.98 ± 68.70	3990.88 ± 62.36	3355.84 ± 49.80	3190.36 ± 183.16	3272.96 ± 553.04

Table 4. *Cont.*

	Yeast 1	Yeast 2	Yeast 3	Yeast 4	Spontaneous
Terpenoids					
cis-Linaloloxide	0.09 ± 0.01	0.09 ± 0.03	0.07 ± 0.04	0.04 ± 0.01	0.06 ± 0.01
trans-Linaloloxide	0.24 ± 0.01	0.06 ± 0.02	0.05 ± 0.01	0.03 ± 0.00	0.09 ± 0.00
Linalool	6.45 ± 0.63	12.31 ± 0.06	10.42 ± 0.35	10.16 ± 0.02	8.81 ± 0.21
Geraniol	2.81 ± 0.11	4.25 ± 0.16	3.03 ± 0.06	3.21 ± 0.21	3.02 ± 0.49
α-Terpineol	5.21 ± 0.01	10.13 ± 0.52	6.46 ± 0.33	4.42 ± 0.26	3.98 ± 0.16
β-Citronellol	14.28 ± 0.25	10.02 ± 0.18	9.40 ± 0.81	9.18 ± 0.39	9.02 ± 1.48
α-Phellandrene	5.70 ± 0.23	10.37 ± 0.98	5.92 ± 0.57	6.62 ± 0.26	6.70 ± 0.28
α-Terpinen	0.20 ± 0.00	0.68 ± 0.54	0.20 ± 0.00	0.19 ± 0.01	0.20 ± 0.01
β-Myrcene	7.31 ± 0.43	13.46 ± 1.27	7.69 ± 0.74	8.59 ± 0.34	8.70 ± 0.36
Limonene	1.38 ± 0.05	2.51 ± 0.14	1.36 ± 0.11	1.46 ± 0.10	1.26 ± 0.02
1,4-Cineole	0.00 ± 0.00	0.05 ± 0.00	0.08 ± 0.00	0.06 ± 0.01	0.06 ± 0.00
1,8-Cineole	0.12 ± 0.02	0.19 ± 0.01	0.09 ± 0.00	0.09 ± 0.01	0.08 ± 0.01
p-Cymene	0.39 ± 0.04	0.55 ± 0.01	0.45 ± 0.03	0.40 ± 0.03	0.34 ± 0.10
Terpinolene	0.84 ± 0.05	1.64 ± 0.23	0.84 ± 0.05	0.85 ± 0.04	0.75 ± 0.05
Terpinen-4-ol	1.23 ± 0.04	1.17 ± 0.06	2.44 ± 0.20	1.59 ± 0.18	1.16 ± 1.12
Norisoprenoids					
β-Damascenone	3.34 ± 0.13	5.00 ± 0.26	2.81 ± 0.16	2.89 ± 0.13	2.23 ± 0.20
3-Hydroxy-β-damascone	0.31 ± 0.04	0.13 ± 0.02	0.16 ± 0.01	0.16 ± 0.01	0.29 ± 0.08
Vitispirane	1.83 ± 0.08	4.01 ± 0.18	1.59 ± 0.07	1.56 ± 0.12	0.76 ± 0.21
TPB	0.02 ± 0.00	0.06 ± 0.01	0.02 ± 0.00	0.02 ± 0.00	0.01 ± 0.00
TDN	0.32 ± 0.02	0.75 ± 0.06	0.24 ± 0.03	0.23 ± 0.03	0.13 ± 0.00
Benzenoids and others					
Benzyl alcohol	199.41 ± 1.32	223.31 ± 21.96	193.43 ± 0.15	233.75 ± 2.73	256.88 ± 5.20
Vanillin	4.97 ± 0.02	4.97 ± 0.06	4.90 ± 0.19	4.87 ± 0.15	4.88 ± 0.14
Ethyl vanillate	135.24 ± 4.24	143.78 ± 4.97	137.46 ± 5.40	140.60 ± 7.06	143.45 ± 4.02
Methyl vanillate	4.71 ± 0.10	6.05 ± 0.60	5.92 ± 0.39	6.00 ± 0.80	4.39 ± 0.20
Benzaldehyde	15.42 ± 0.74	15.45 ± 0.39	15.60 ± 0.86	16.59 ± 1.25	15.47 ± 1.19
Eugenol	6.55 ± 0.35	6.28 ± 0.01	5.46 ± 0.01	5.14 ± 0.17	6.00 ± 0.02
Methyl salicylate	3.64 ± 0.07	5.73 ± 0.09	2.32 ± 0.05	2.99 ± 0.08	2.35 ± 0.17
2,6-Dimethoxyphenol	6.51 ± 0.01	6.57 ± 0.69	5.51 ± 0.16	5.30 ± 0.45	5.78 ± 0.89
Furfural	1.68 ± 0.11	2.06 ± 0.35	1.54 ± 0.08	1.43 ± 0.12	1.53 ± 0.13
γ-Decalactone	2.60 ± 0.14	2.25 ± 0.01	2.36 ± 0.15	2.44 ± 0.07	2.26 ± 0.01
δ-Decalactone	25.14 ± 8.63	21.78 ± 2.94	25.17 ± 3.28	23.91 ± 0.01	23.58 ± 1.71

Nd means not detected. TPB is short for (E)-1-(2,3,6-Trimethylphenyl)-buta-1,3-diene. TDN is short for 1,6,-trimethyl-1,2-dihydronapthalene.

Table 5. Concentration (μg/L) of volatile compounds of Area 1 Corvinone withered wines.

	Yeast 1	Yeast 2	Yeast 3	Yeast 4	Spontaneous
	mean ± sd	mean ± sd	mean ± sd	mean ± sd	mean ± sd
Alcohols					
1-Butanol	326.74 ± 172.01	228.80 ± 28.38	158.29 ± 15.66	100.38 ± 4.53	141.27 ± 17.97
2-Butanol	5643.65 ± 610.53	4982.00 ± 428.99	5859.68 ± 54.54	3285.67 ± 159.30	10321.02 ± 654.72
1-Pentanol	48.14 ± 0.59	43.09 ± 3.20	46.67 ± 7.15	59.99 ± 4.31	42.16 ± 1.04
Isoamyl alcohol (mg/L)	254.28 ± 83.86	27.14 ± 10.10	28.44 ± 27.64	200.64 ± 13.64	179.62 ± 7.03
Phenylethyl alcohol (mg/L)	21.33 ± 0.99	19.42 ± 0.03	19.27 ± 2.09	20.95 ± 0.99	14671 ± 1.35
Methionol	221.98 ± 5.51	135.32 ± 28.73	222.60 ± 31.58	389.66 ± 73.67	51.24 ± 15.38
C_6 alcohols					
1-Hexanol	2020.95 ± 110.53	1940.31 ± 3.38	1737.24 ± 135.60	1701.42 ± 125.04	1710.03 ± 34.29
trans-3-Hexen-1-ol	27.90 ± 1.53	27.90 ± 0.91	29.85 ± 0.43	25.66 ± 1.57	17.26 ± 3.01
cis-3-Hexen-1-ol	17.83 ± 0.05	20.29 ± 1.92	19.81 ± 1.68	18.76 ± 1.36	19.12 ± 4.82
cis-2-Hexen-1-ol	16.54 ± 0.03	15.05 ± 0.83	14.73 ± 0.55	16.21 ± 0.88	15.78 ± 1.07
Acetate esters					
Isoamyl acetate	382.14 ± 9.68	707.32 ± 20.52	482.77 ± 29.00	258.90 ± 31.35	547.90 ± 4.36
n-Hexyl acetate	17.77 ± 0.84	14.17 ± 1.07	8.98 ± 0.74	12.35 ± 2.60	11.60 ± 0.49
2-Phenylethyl acetate	35.58 ± 0.69	29.21 ± 0.07	25.74 ± 1.08	24.61 ± 1.47	42.36 ± 1.78
Ethyl acetate (mg/L)	64.37 ± 10.73	101.32 ± 11.63	62.12 ± 15.22	44.99 ± 1.17	159.83 ± 0.86

Table 5. *Cont.*

	Yeast 1	Yeast 2	Yeast 3	Yeast 4	Spontaneous
Branched-chain fatty acids ethyl esters					
Ethyl 2-methylbutanoate	1.80 ± 0.35	3.57 ± 0.21	2.46 ± 0.31	1.21 ± 0.21	0.89 ± 0.16
Ethyl 3-methylbutanoate	3.13 ± 0.01	3.10 ± 0.16	3.22 ± 0.13	3.06 ± 0.09	3.19 ± 0.10
Fatty acids ethyl esters					
Ethyl butanoate	159.71 ± 5.76	206.20 ± 1.07	184.08 ± 12.60	168.03 ± 9.72	125.35 ± 16.24
Ethyl hexanoate	288.88 ± 34.73	441.20 ± 3.88	371.86 ± 22.02	294.37 ± 1.03	136.72 ± 21.62
Ethyl octanoate	130.27 ± 11.62	204.82 ± 17.71	182.78 ± 16.91	117.46 ± 7.09	63.44 ± 3.55
Ethyl lactate	467.73 ± 59.32	685.54 ± 32.99	502.66 ± 4.44	415.45 ± 8.14	1047.19 ± 10.42
Ethyl decanoate	31.85 ± 2.35	45.66 ± 2.81	54.99 ± 2.70	29.99 ± 3.01	6.92 ± 1.84
Other esters					
Ethyl 3-hydroxybutanoate	219.48 ± 14.79	291.85 ± 20.39	373.57 ± 19.01	169.63 ± 3.90	36.88 ± 0.49
Ethyl 2-hydroxyhexanoate	0.40 ± 0.10	1.36 ± 0.03	1.22 ± 0.02	0.66 ± 0.03	0.22 ± 0.09
Fatty acids					
3-Methylbutanoic acid	323.93 ± 49.67	364.47 ± 17.78	486.21 ± 29.49	253.57 ± 14.10	363.79 ± 279.85
Hexanoic acid	1882.76 ± 89.07	2938.93 ± 68.25	2359.40 ± 124.09	1986.87 ± 33.29	1022.20 ± 80.26
Octanoic acid	3576.79 ± 173.28	4432.38 ± 27.19	4197.55 ± 25.65	3798.66 ± 22.14	1735.40 ± 831.93
Terpenoids					
cis-Linaloloxide	0.11 ± 0.08	0.06 ± 0.02	0.08 ± 0.04	0.05 ± 0.02	0.95 ± 1.24
trans-Linaloloxide	0.07 ± 0.01	0.03 ± 0.01	0.08 ± 0.08	0.04 ± 0.01	0.32 ± 0.26
Linalool	4.36 ± 0.19	5.06 ± 0.50	4.49 ± 0.30	1.55 ± 0.07	5.05 ± 0.25
Geraniol	0.83 ± 0.25	1.14 ± 0.26	1.15 ± 0.21	1.32 ± 0.17	1.10 ± 0.17
α-Terpineol	3.19 ± 0.57	3.57 ± 0.53	4.06 ± 0.67	3.37 ± 0.04	3.39 ± 0.11
β-Citronellol	5.62 ± 1.79	4.58 ± 0.16	7.25 ± 0.38	3.68 ± 0.04	3.53 ± 0.64
α-Phellandrene	1.08 ± 0.01	1.38 ± 0.07	1.74 ± 0.17	0.41 ± 0.01	2.19 ± 0.42
α-Terpinen	0.06 ± 0.01	0.09 ± 0.01	0.09 ± 0.02	0.22 ± 0.02	0.10 ± 0.01
β-Myrcene	1.40 ± 0.03	1.80 ± 0.09	2.26 ± 0.23	2.64 ± 0.11	2.85 ± 0.54
Limonene	0.40 ± 0.01	0.46 ± 0.02	0.61 ± 0.11	0.85 ± 0.04	0.69 ± 0.08
1,4-Cineole	0.03 ± 0.01	0.02 ± 0.00	nd	nd	nd
1,8-Cineole	0.09 ± 0.01	0.08 ± 0.01	nd	nd	0.09 ± 0.01
p-Cymene	0.16 ± 0.02	0.24 ± 0.01	0.22 ± 0.03	0.26 ± 0.03	0.25 ± 0.03
Terpinolene	0.27 ± 0.03	0.33 ± 0.01	0.35 ± 0.02	0.53 ± 0.02	0.36 ± 0.02
Terpinen-4-ol	0.11 ± 0.01	0.12 ± 0.01	0.82 ± 0.02	0.10 ± 0.00	0.26 ± 0.06
Norisoprenoids					
β-Damascenone	5.46 ± 0.07	5.45 ± 0.92	10.30 ± 0.28	7.93 ± 1.32	9.07 ± 0.06
3-Hydroxy-β-damascone	0.18 ± 0.02	0.16 ± 0.05	0.26 ± 0.01	0.24 ± 0.01	0.14 ± 0.01
Vitispirane	9.77 ± 0.66	10.79 ± 1.84	11.99 ± 3.46	16.45 ± 0.47	13.37 ± 0.69
TPB	0.07 ± 0.01	0.08 ± 0.01	0.18 ± 0.01	0.17 ± 0.01	0.12 ± 0.02
TDN	1.81 ± 0.01	1.76 ± 0.37	3.69 ± 0.21	3.70 ± 0.01	2.78 ± 0.42
Benzenoids and others					
Benzyl alcohol	45.93 ± 4.19	35.16 ± 1.95	44.49 ± 2.60	56.89 ± 5.24	72.90 ± 2.55
Vanillin	6.06 ± 0.25	5.94 ± 0.22	5.94 ± 0.06	5.88 ± 0.17	6.04 ± 0.08
Ethyl vanillate	182.83 ± 18.82	172.96 ± 1.77	189.04 ± 7.86	170.44 ± 4.26	191.00 ± 20.58
Methyl vanillate	11.69 ± 0.74	9.33 ± 0.24	14.28 ± 1.34	11.61 ± 0.53	12.30 ± 0.12
Benzaldehyde	14.50 ± 0.22	14.30 ± 0.19	15.03 ± 0.21	14.75 ± 0.00	14.60 ± 0.14
Eugenol	1.90 ± 0.01	1.88 ± 0.13	2.01 ± 0.28	1.88 ± 0.11	1.88 ± 0.11
Methyl salicylate	1.52 ± 0.18	1.57 ± 0.25	3.92 ± 0.01	3.05 ± 0.06	2.48 ± 0.12
2,6-Dimethoxyphenol	5.28 ± 0.06	5.07 ± 0.12	4.83 ± 0.37	5.04 ± 0.39	6.11 ± 0.27
Furfural	0.67 ± 0.07	1.04 ± 0.72	0.97 ± 0.10	0.97 ± 0.04	0.63 ± 0.15
γ-Decalactone	2.47 ± 0.17	3.30 ± 0.28	3.09 ± 0.32	2.32 ± 0.11	2.34 ± 0.02
δ-Decalactone	17.63 ± 3.57	23.47 ± 2.17	20.19 ± 1.21	18.71 ± 0.62	14.87 ± 0.81

Nd means not detected. TPB is short for (E)-1-(2,3,6-Trimethylphenyl)-buta-1,3-diene. TDN is short for 1,6,-trimethyl-1,2-dihydronapthalene.

Table 6. Concentration (μg/L) of volatile compounds of Area 2 Corvinone wines.

	Yeast 1	Yeast 2	Yeast 3	Yeast 4	Spontaneous
	mean ± sd	mean ± sd	mean ± sd	mean ± sd	mean ± sd
Alcohols					
1-Butanol	435.89 ± 5.62	246.54 ± 7.13	202.54 ± 3.61	161.62 ± 28.98	203.59 ± 32.91
2-Butanol	3865.04 ± 40.39	4224.78 ± 295.96	4888.37 ± 123.11	4074.09 ± 56.97	4692.72 ± 419.96
1-Pentanol	69.68 ± 10.63	53.39 ± 3.74	51.05 ± 1.63	40.72 ± 1.02	36.36 ± 1.58
Isoamyl alcohol (mg/L)	25.80 ± 8.01	32.46 ± 25010	266.04 ± 4.14	312.92 ± 9.35	218.75 ± 2.01
Phenylethyl alcohol (mg/L)	21.67 ± 0.30	26.48 ± 2.34	20.01 ± 0.02	24.69 ± 1.02	15.01 ± 0.06
Methionol	116.97 ± 4.76	217.18 ± 60.02	129.93 ± 9.23	259.07 ± 18.89	69.61 ± 2.59
C_6 alcohols					
1-Hexanol	1149.65 ± 29.59	1424.44 ± 70.95	1054.97 ± 33.81	1154.77 ± 40.13	1189.48 ± 53.31
trans-3-Hexen-1-ol	15.01 ± 1.16	19.05 ± 1.02	13.65 ± 0.96	18.31 ± 0.26	13.40 ± 1.28
cis-3-Hexen-1-ol	11.12 ± 0.28	15.55 ± 0.16	11.72 ± 0.59	15.26 ± 0.30	13.78 ± 1.94
cis-2-Hexen-1-ol	16.00 ± 0.44	15.68 ± 0.33	16.11 ± 0.35	17.28 ± 0.94	15.07 ± 0.94
Acetate esters					
Isoamyl acetate	702.55 ± 31.55	689.58 ± 45.25	450.43 ± 15.25	373.92 ± 23.64	708.08 ± 37.99
n-Hexyl acetate	10.75 ± 1.54	12.18 ± 3.84	8.81 ± 2.03	17.77 ± 0.24	14.02 ± 2.14
2-Phenylethyl acetate	26.14 ± 2.51	26.53 ± 2.86	14.21 ± 1.85	14.32 ± 2.01	32.92 ± 2.21
Ethyl acetate (mg/L)	57.70 ± 1.62	70.32 ± 9.73	69.85 ± 1.60	54.02 ± 6.67	158.50 ± 2.59
Branched-chain fatty acids ethyl esters					
Ethyl 2-methylbutanoate	1.83 ± 0.04	2.53 ± 0.04	2.31 ± 0.14	2.93 ± 0.09	1.12 ± 0.08
Ethyl 3-methylbutanoate	3.11 ± 0.14	3.10 ± 0.13	3.21 ± 0.16	2.95 ± 0.08	3.10 ± 0.18
Fatty acids ethyl esters					
Ethyl butanoate	267.28 ± 27.74	366.95 ± 11.50	201.25 ± 3.67	241.66 ± 12.76	262.34 ± 21.33
Ethyl hexanoate	415.84 ± 14.28	567.81 ± 45.25	358.27 ± 11.13	439.93 ± 23.36	364.84 ± 50.11
Ethyl octanoate	221.11 ± 6.07	320.36 ± 68.82	236.13 ± 6.18	282.17 ± 7.74	198.47 ± 11.00
Ethyl lactate	226.73 ± 7.34	443.77 ± 29.85	206.46 ± 8.03	277.63 ± 36.13	531.84 ± 41.32
Ethyl decanoate	51.72 ± 7.74	85.86 ± 5.85	67.95 ± 2.79	75.81 ± 10.45	48.74 ± 6.30
Other esters					
Ethyl 3-hydroxybutanoate	198.08 ± 5.95	266.81 ± 20.89	215.93 ± 5.90	168.82 ± 12.93	128.57 ± 28.39
Ethyl 2-hydroxyhexanoate	0.31 ± 0.04	1.42 ± 0.04	0.66 ± 0.06	0.84 ± 0.06	0.33 ± 0.04
Fatty acids					
3-Methylbutanoic acid	374.47 ± 10.27	432.56 ± 57.91	364.05 ± 17.82	459.10 ± 2.18	282.97 ± 14.81
Hexanoic acid	2194.09 ± 377.72	3172.54 ± 401.13	2056.21 ± 65.63	2299.39 ± 149.00	1868.12 ± 27.08
Octanoic acid	4021.06 ± 130.90	5095.48 ± 359.10	3948.40 ± 101.82	4166.92 ± 87.94	3786.14 ± 46.03
Terpenoids					
cis-Linaloloxide	0.07 ± 0.02	0.11 ± 0.06	0.10 ± 0.01	0.04 ± 0.01	0.08 ± 0.01
trans-Linaloloxide	0.08 ± 0.07	0.04 ± 0.00	0.03 ± 0.00	0.08 ± 0.00	0.05 ± 0.01
Linalool	5.34 ± 0.19	6.44 ± 0.05	6.30 ± 0.29	6.52 ± 0.52	5.55 ± 0.66
Geraniol	1.09 ± 0.20	1.67 ± 0.24	1.75 ± 0.21	1.75 ± 0.26	1.68 ± 0.30
α-Terpineol	3.08 ± 0.07	4.18 ± 0.11	4.30 ± 0.32	3.51 ± 0.13	3.41 ± 0.30
β-Citronellol	10.93 ± 0.36	6.57 ± 0.17	7.82 ± 0.84	7.94 ± 0.42	6.92 ± 0.28
α-Phellandrene	2.90 ± 0.16	2.79 ± 0.04	3.07 ± 0.18	3.45 ± 0.35	2.97 ± 0.24
α-Terpinen	0.09 ± 0.01	0.14 ± 0.12	0.05 ± 0.00	0.06 ± 0.01	0.05 ± 0.01
β-Myrcene	3.76 ± 0.21	3.68 ± 0.04	3.98 ± 0.24	4.48 ± 0.45	3.89 ± 0.27
Limonene	0.68 ± 0.05	0.66 ± 0.08	0.71 ± 0.04	0.72 ± 0.04	0.67 ± 0.03
1,4-Cineole	0.04 ± 0.00	nd	nd	nd	nd
1,8-Cineole	0.09 ± 0.00	0.11 ± 0.01	0.10 ± 0.01	0.08 ± 0.01	0.09 ± 0.01
p-Cymene	0.24 ± 0.01	0.16 ± 0.01	0.15 ± 0.00	0.15 ± 0.00	0.17 ± 0.02
Terpinolene	0.33 ± 0.05	0.22 ± 0.03	0.31 ± 0.01	0.29 ± 0.02	0.27 ± 0.03
Terpinen-4-ol	1.59 ± 0.03	0.05 ± 0.01	0.06 ± 0.01	0.05 ± 0.00	0.06 ± 0.01
Norisoprenoids					
β-Damascenone	3.19 ± 0.08	2.77 ± 0.15	2.87 ± 0.04	3.03 ± 0.25	2.54 ± 0.23
3-Hydroxy-β-damascone	0.12 ± 0.01	0.15 ± 0.00	0.10 ± 0.01	0.11 ± 0.01	0.15 ± 0.05
Vitispirane	3.56 ± 0.28	4.42 ± 0.14	3.43 ± 0.14	3.54 ± 0.17	3.44 ± 0.23
TPB	0.02 ± 0.00	0.02 ± 0.00	0.02 ± 0.00	0.02 ± 0.00	0.02 ± 0.00
TDN	0.33 ± 0.04	0.34 ± 0.03	0.24 ± 0.01	0.26 ± 0.01	0.26 ± 0.02

Table 6. *Cont.*

	Yeast 1	Yeast 2	Yeast 3	Yeast 4	Spontaneous
Benzenoids and others					
Benzyl alcohol	39.17 ± 0.30	34.19 ± 1.61	31.11 ± 3.26	52.14 ± 5.28	56.54 ± 0.81
Vanillin	5.45 ± 0.07	5.54 ± 0.16	5.49 ± 0.21	5.60 ± 0.21	5.50 ± 0.14
Ethyl vanillate	177.78 ± 4.48	185.85 ± 22.39	145.92 ± 19.79	184.34 ± 21.53	180.38 ± 33.66
Methyl vanillate	10.56 ± 0.22	13.24 ± 0.30	11.47 ± 0.72	12.80 ± 1.77	12.04 ± 1.13
Benzaldehyde	14.50 ± 0.09	15.03 ± 0.69	14.19 ± 0.49	14.84 ± 0.24	14.80 ± 0.11
Eugenol	1.26 ± 0.01	1.26 ± 0.01	1.16 ± 0.06	1.26 ± 0.01	1.68 ± 0.01
Methyl salicylate	0.64 ± 0.07	0.77 ± 0.04	0.63 ± 0.13	0.64 ± 0.08	0.37 ± 0.02
2,6-Dimethoxyphenol	6.70 ± 0.42	5.66 ± 0.53	5.11 ± 0.65	6.98 ± 0.03	5.99 ± 0.37
Furfural	1.79 ± 0.08	2.10 ± 0.10	1.62 ± 0.12	1.86 ± 0.19	1.35 ± 0.19
γ-Decalactone	2.25 ± 0.00	3.20 ± 0.16	2.47 ± 0.30	3.02 ± 0.11	2.32 ± 0.08
δ-Decalactone	22.06 ± 0.52	28.68 ± 4.83	16.57 ± 2.34	28.13 ± 2.51	23.11 ± 2.44

Nd means not detected. TPB is short for (E)-1-(2,3,6-Trimethylphenyl)-buta-1,3-diene. TDN is short for 1,6,-trimethyl-1,2-dihydronapthalene.

The influence of spontaneous fermentation was assessed by comparing, for each compound, average concentrations of all inoculated fermentations with that of spontaneous fermentation ($\alpha = 0.05$) (Table S4). In Corvina and Corvinone wines from Area 1, twenty-eight and twenty-one compounds were found to be significantly affected, respectively, while in Area 2 fourteen and seventeen compounds were found to be significantly affected, respectively.

An overview of the relative influence of the different variables was obtained by Principal Component Analysis (PCA). In the case of Corvina wines (Figure 1a), 49.45% of the total variance could be explained with the first two principal components (PCs). PC1, accounting for 27.2% of the variance, was associated with grape origin, while PC2, accounting for 22.24% of the total variance, differentiated yeast strain/inoculation strategy. Separation along PC1 was mostly associated with differences in the concentration of compounds such as terpene alcohols, norisoprenoids, benzenoids and branched chain fatty acid esters, whereas along PC2, separation was mostly associated with ethyl esters, acetate esters, ethyl acetate, and higher alcohols. In the case of Corvinone wines (Figure 1b), 50.58% of the total variance was explained with the first two principal components. Additionally in this case, PC1, accounting for 33.31% of the total variance, was associated with grape origin, whereas PC2, accounting for 17.27% of the total variance, was associated with yeast strain/inoculation. Separation along PC1 was mostly associated with differences in the concentration of compounds such as terpene alcohols, norisoprenoids, benzenoids, ethyl esters, fatty acids, and higher alcohols, whereas along PC2, separation was mostly associated with ethyl acetate and methionol.

Among the compounds accounting for differences associated with grape origin, terpenes were found to be largely discriminant, with Area 2 wines showing higher content. The most abundant terpenes were β-citronellol, linalool, and in the case of Corvina, β-myrcene. The fact that terpenes were distinctive of grape origin is in agreement with other reports indicating these compounds as good markers of terroir influence [4,37–41]. Terpenes have also been indicated as possible varietal markers of Corvina and Corvinone wines compared with other red wines [7], so that their variations in response to grape origin appear of particular interest in the context of Valpolicella wines. Differences in wine terpene content have previously been related to the degree of grape maturity, as grape terpenes tend to increase with ripening [31,42,43]. In the present study, Area 2 had significantly higher sugar levels at crush, which appeared to be associated with an increased content of terpenes such as linalool, α-terpineol, and myrcene in the wines. It has to be considered, however, that differences in sugar content at crush were mostly due not to maturity in itself, but to withering, which can induce important modification in grape terpene content [27]. Moreover, citronellol, the most abundant among the terpene measured, showed a partially opposite behaviour, with significantly higher content in Area 1 wines for Corvina. In non-aromatic grapes such as Corvina and Corvinone, formation of compounds such as linalool is primarily due to release from glycosidic precursors [4,44], whereas citronellol mostly arises from yeast-driven reduction of the geraniol released from precursors [44]. In this sense, citronellol was also the terpene showing the largest variations due to yeast strain, supporting the view that yeast action was a main factor to the release of this compound.

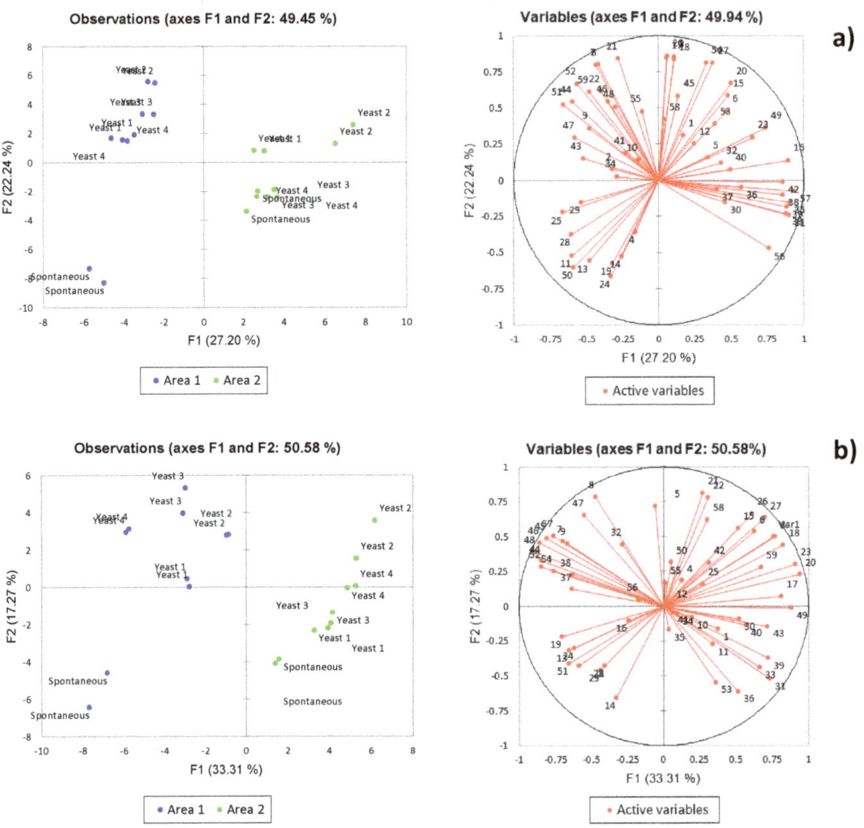

Figure 1. PCA of (**a**) Corvina and (**b**) Corvinone wines. Variables plot numbers correspond to: (1) 1-Butanol, (2) Butanol, (3) Isoamyl alcohol, (4) 1-Pentanol, (5) Methionol, (6) Phenylethyl alcohol, (7) 1-Hexanol, (8) *trans*-3-Hexen-1-ol, (9) *cis*-3-Hexen-1-ol, (10) *cis*-2-Hexen-1-ol, (11) Isoamyl acetate, (12) n-Hexyl acetate, (13) 2-Phenylethyl acetate, (14) Ethyl acetate, (15) Ethyl-2-methylbutanoate, (16) Ethyl 3-methylbutanoate, (17) Ethyl butanoate, (18) Ethyl hexanoate, (19) Ethyl lactate, (20) Ethyl octanoate, (21) Ethyl 3-hydroxybutanoate, (22) Ethyl 2-hydroxyhexanoate, 23) Ethyl decanoate, (24) Acetic acid, (25) 3-Methylbutanoic acid, (26) Hexanoic acid, (27) Octanoic acid, (28) *cis*-Linaloloxide, (29) *trans*-Linaloloxide, (30) 3-carene, (31) α-Phellandrene, (32) α-Terpinene, (33) β-Myrcene, (34) 1,4-Cineole, (35) Limonene, (36) 1,8-cineole, (37) p-Cymene, 38) Terpinolene, (39) Linalool, (40) Geraniol, (41) Terpinen-4-ol, (42) α-Terpineol, (43) β-Citronellol, (44) β-Damascenone, (45) TPB (E)-1-(2,3,6-Trimethylphenyl)-buta-1,3-diene), (46) TDN (1,6,-trimethyl-1,2-dihydronapthalene), (47) 3-Hydroxy-β-damascone, (48) Vitispirane, (49) Furfural, (50) Benzaldehyde, (51) Benzyl alcohol, (52) Eugenol, (53) 2,6-Dimethoxyphenol, (54) Vanillin, (55) Methyl vanillate, (56) Ethyl vanillate, (57) Methyl salicylate, (58) γ-Decalactone, (59) δ-Decalactone.

Similar to terpenes, norisoprenoids were associated with different grape origins, although in this case higher levels occurred in Area 1 wines. This further confirms the lack of a clear direct association between grape sugar content and wine terpenoid content, as norisoprenoids would also be expected to increase with higher maturity. It has been previously shown that, while wine terpene content tends to decrease with withering, that of norisoprenoids tends to increase [6], so that the outcomes of withering are only partly similar to those of prolonged ripening.

Among norisoprenoids, major variations were observed for the potent odorant β-damascenone, which in Area 1 samples attained significantly higher concentrations. As in the case of citronellol, this compound arises from a complex mechanism in which grape precursor content as well as yeast ability to express both glycosidase and reductase activities are expected to play a role [45].

Esters were another group of volatile compounds associated with significant compositional differences. In agreement with the well-known influence of yeast strain on the production of these compounds during fermentation [14], a major influence of yeast strain and spontaneous fermentation was observed (Figure 1a,b). However, depending on ester chemical class, the relationship between inoculated/spontaneous fermentation, yeast strain and ester production was also affected by grape terroir of origin, and by grape variety. This can be seen in Figure 2, a heat map with the four main esters classes analyzed in this study.

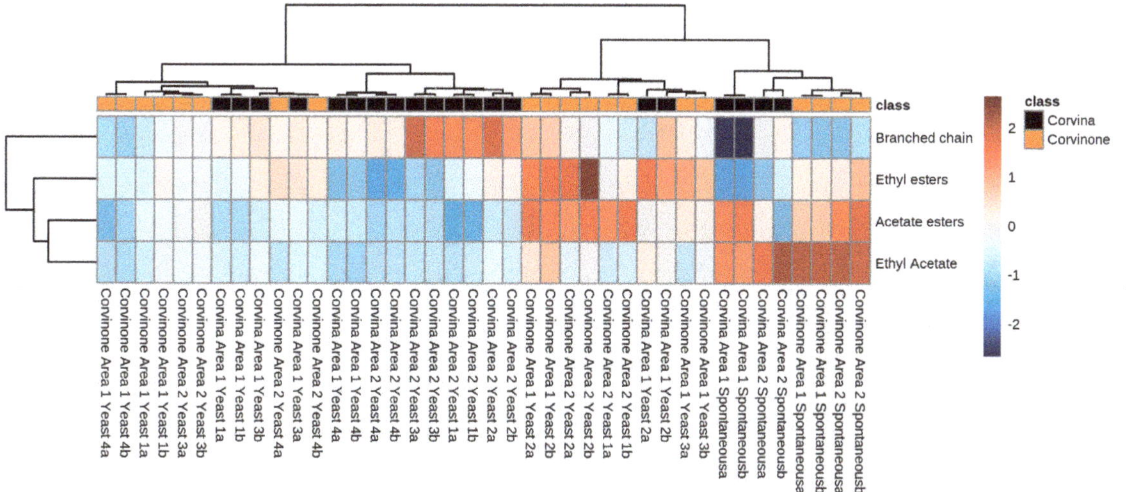

Figure 2. Heat map of fermentative esters in all studied wines.

Wines from spontaneous fermentations were well differentiated from those obtained by inoculation with commercial yeasts, regardless of grape variety and area of origin. This was due to an increased concentration of ethyl acetate and acetate esters and low content of all other esters. Most Corvinone wines were grouped together in a second cluster characterized by a higher content of ethyl and acetate esters, and this cluster also included most of the wines obtained with Yeast 2. Conversely, the majority of Corvina wines from Area 2 were in a third cluster with higher content of branched chain fatty acid esters and lower content of ethyl esters and acetates. Finally, a fourth cluster including Corvina and Corvinone wines fermented with yeasts 1, 3 and 4, were characterised by intermediate values for all the esters.

Clearly, ethyl acetate production during fermentation introduced major differences in the wine ester profile. This compound is a common constituent of fermented beverages, although winemakers are commonly interested in limiting its presence due to the distinctive nail polish aroma that it can impart to wines at high concentrations. Biosynthesis of ethyl acetate in yeast is associated with the maintenance of adequate intracellular levels of CoA, which can be restored from acetyl-CoA by an appropriate alcohol acetyltransferase in the presence of ethanol, with consequent formation of ethyl acetate [46,47]. When high concentrations of both acetic acid and ethanol are present, ethyl acetate can also be formed via ethanol esterification by an esterase [46]. Spontaneous fermentations are typically associated with increased acetic acid and ethyl acetate concentrations, due to the greater proliferation of indigenous non-*Saccharomyces* species [48]. Due to their relatively low ethanol tolerance, non-*Saccharomyces* yeasts tend to proliferate in the early stages of fermentation, while with progressive accumulation of ethanol it is *S. cerevisiae* that eventually takes dominance of the fermentation medium. Therefore, it is often possible that, in the same work environment (e.g., a cellar), a given *S. cerevisiae* strain dominates both spontaneous and inoculated fermentation [48], which could have been the case also in the present study. Nevertheless, the data clearly indicated that the practice of spontaneous fermentation (which also required avoiding the addition of SO_2) was systematically associated with specific volatile profiles that were firstly characterized by higher levels of acetic acid and ethyl acetate. As for inoculated fermentations, the data indicated that the choice of yeast strain could induce significant differences in ethyl acetate levels, which could vary up to approximately 50% for the same must, with Yeast 2 generally producing increased ethyl acetate (Figure 3). However, this same yeast was also characterized by an increased production of fatty acid ethyl esters with pleasant fruity aromas such as ethyl hexanoate and ethyl octanoate, appearing overall as a high ester producing strain.

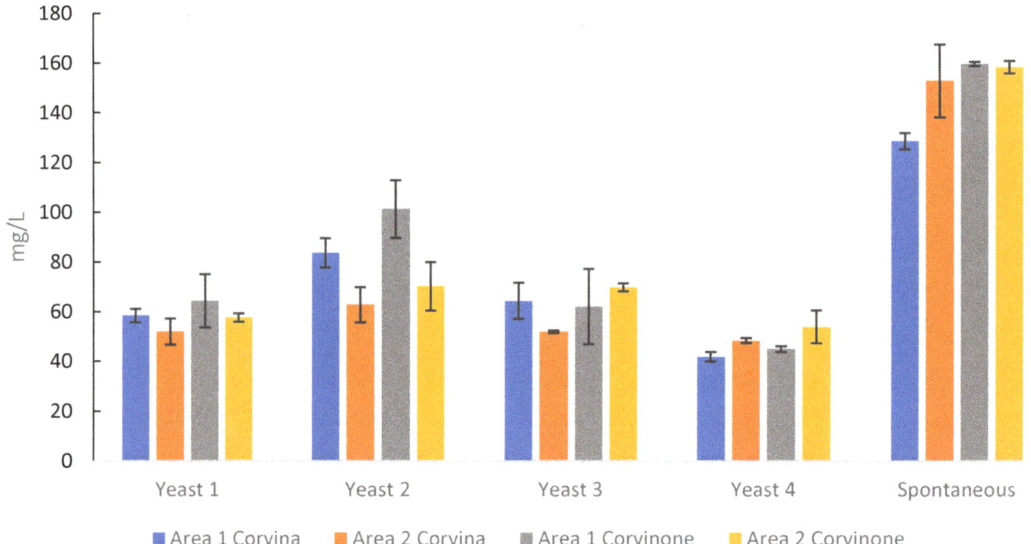

Figure 3. Concentration of ethyl acetate in wines obtained with different commercial strains as well as by spontaneous fermentation. Error bars indicate ± standard deviation.

3.3. Sensory Analysis

In studies in which multiple variables are compared, a comparative assessment of the sensory impact of individual variables can be effectively carried out evaluating the existence of odor similarity patterns that could be associated with one or more of the studied variables [49]. For this purpose, a sorting task was carried out using the samples of the study, and results elaborated by means of HCA are shown in Figure 4. Replicate wines were projected in the same group, meaning that the panel was reproducible and the wine replicates were similar. In the case of Corvina (Figure 4a), wines were clustered in three groups. A first cluster consisted of all inoculated wines from Area 1, a second cluster consisted of all the spontaneous fermentations regardless of grape origin, and finally, a third cluster consisted of all the inoculated wines from Area 2. Additionally, in the case of Corvinone (Figure 4b), three clusters were obtained. The first cluster consisted, again, of all inoculated wines from Area 1, the second cluster consisted of four wines, all from Area 2, fermented with yeast 1 and 4, and finally, the third cluster consisted of the remaining Area 2 wines and all the spontaneous fermentations. It appears therefore clear that, from a sensory perspective, not only grape origin but also the type of fermentation inoculum is strongly associated with perceivable odor differences. Indeed, taking for example the case of Corvina, on the one hand, two main clusters were observed, each containing all the samples of either Area 1 or 2 fermented with commercial strains, which confirmed the primary role of the terroir of grape origin. On the other hand, a third cluster was observed, consisting of all wines obtained by means of spontaneous fermentation, regardless of the terroir of grape origin. The contribution of yeast to the expression of terroir attributes has been investigated by different studies, and it has been often hypothesized that complex microbial consortia such as those associated with spontaneous fermentation might result in the expression of unique sensory features associated with specific vineyard sites [50]. Conversely, in the case of the present study, in particular for Corvina, it was inoculation with commercial strains that allowed differentiation based on grape origin, whereas spontaneous fermentation produced wines with distinctive odor profiles that could be clearly grouped by the panel, regardless of grape origin.

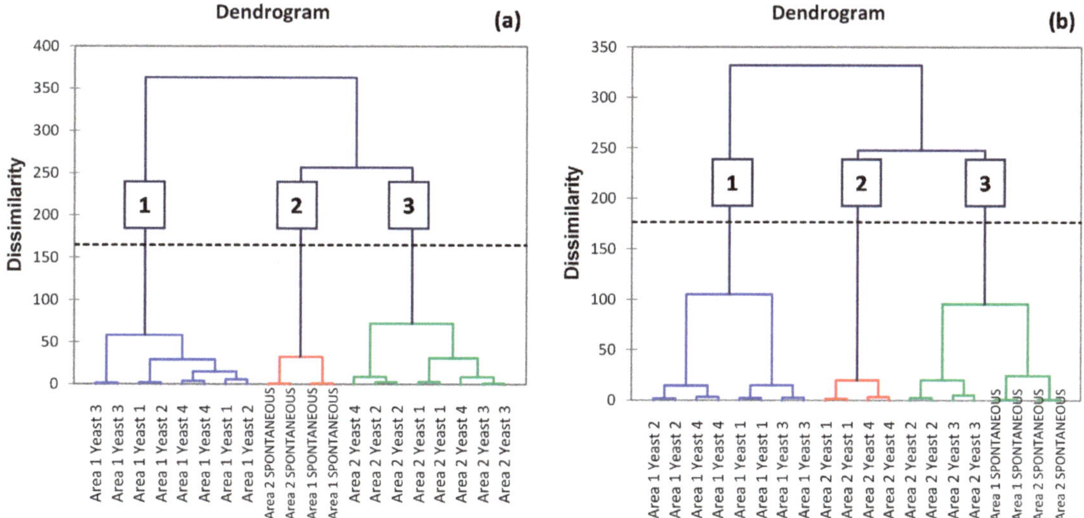

Figure 4. HCA of (**a**) Corvina and (**b**) Corvinone withered grapes wines sorting task data.

In order to identify the volatile compounds most likely contributing to the observed clusters, sorting task data were compared with wines' volatile composition. In the case of Corvina, thirty-five volatile compounds showed a significantly different content between clusters, thirteen in the case of Corvinone (Table S5). Among these, a comparison of compounds with an odor activity value (OAV) higher than one, which are expected to contribute more prominently to wine aroma, was carried out (Figures 5 and 6).

The data obtained indicated that, both for Corvina and Corvinone, the cluster associated with spontaneous fermentation (namely clusters 2 and 3, respectively) was characterized by significantly higher levels of acetic acid and ethyl acetate. This appears in agreement with the observation that the initial stages of spontaneous fermentation are typically associated with increased proliferation of non-*Saccharomyces* yeasts, some of which are characterized by high production of ethyl acetate and acetic acid [51,52]. Additionally, the particularly high levels of these compounds may also be due to a response to osmotic stress [29] from the high glucose and fructose content (up to almost 300 g/L) following the withering treatment. Ethyl acetate is often described as having a nail lacquer odor, whereas acetic acid is the main odor compound of vinegar. These compounds are likely to contribute to the specific odor character of wines from spontaneous fermentation, also in combination with the fact that the levels of other compounds such as norisoprenoids and ethyl esters were generally lower in samples from spontaneous fermentations.

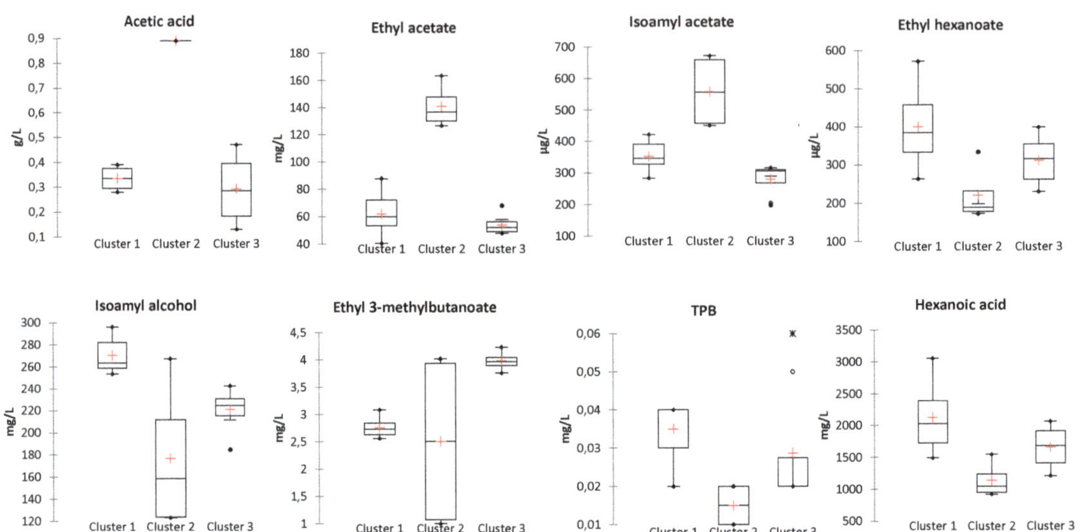

Figure 5. Box plots of significantly different compounds (Kruskal–Wallis, α = 0.05) between sensory clusters of Corvina wines.

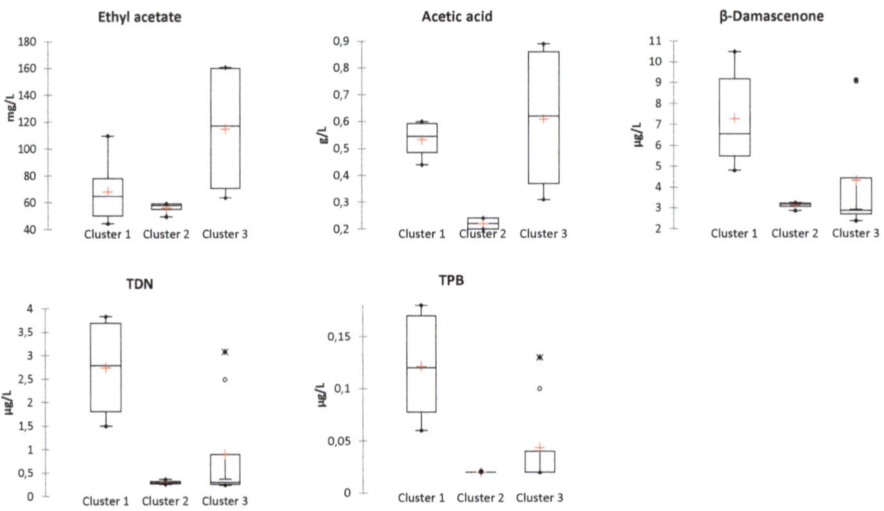

Figure 6. Box plots of significantly different compounds (Kruskal–Wallis, α = 0.05) between sensory clusters of Corvinone wines.

Further interesting insights were obtained concerning the contribution of other aroma compounds to the odor profile of the other clusters observed. In the case of Corvina (Figure 5), wines from cluster 1, all inoculated from Area 1, compared with cluster 3, all inoculated from Area 2, showed a higher content of ethyl butanoate and ethyl hexanoate (fruity), TPB ((E)-1-(2,3,6-Trimethylphenyl)-buta-1,3-diene)(tobacco), TDN (1,6,-trimethyl-1,2-dihydronapthalene) (kerosene), β-damascenone (quince), eugenol (cloves) and hexanoic acid (rancid). Concerning Corvinone, wines from cluster 1 all inoculated from Area 1, showed higher levels of norisoprenoids and eugenol (Figure 6), whereas wines of cluster

2 (Area 2 yeast 1 and 4) were characterized by low levels of acetic acid, below the odor threshold. Although these compounds were present in concentrations exceeding their odor threshold, at the levels at which they were detected they are not expected to specifically impart their odor to the wines of each cluster, but mostly to act synergistically to determine the overall sensory nuances perceived during tasting [11].

4. Conclusions

In conclusion, most aroma-related volatile metabolites of Corvina and Corvinone wines from withered grapes were affected by grape terroir of origin, highlighting the central role of grape selection for obtaining defined Amarone aroma profiles. Sensory analysis confirmed this observation, since grape origin was the major driver of the sensory differences observed. Nevertheless, when spontaneous fermentation was applied, wine sensory characteristics associated with either grape terroir or variety were less prominent, and this was due to increased levels of ethyl acetate and acetic acid. Other volatile compounds, in particular, norisoprenoids and ethyl esters, contributed to expressing sensory diversity associated with either yeast strain or grape terroir of origin. Additional sensory work in determining the actual contribution of these compounds to Amarone aroma is needed.

The results of this study represent a first attempt to classify the relative contribution of different industry-relevant variables to the management of the Amarone aroma profile. Further work is needed to address more specifically the technological factors determining some of the behaviours observed, in particular, concerning yeast ethyl acetate metabolism. This will enable winemakers to rationalize production choices towards specific sensory quality objectives.

Limitation of the study: Equipment: The laboratory did not have a sensory room equipped with individual boots. Time and method: Identification of the bacterial population in spontaneous fermentations would have been interesting information, even if not indispensable to the study.

Supplementary Materials: The following are available online at https://www.mdpi.com/article/10.3390/foods10102474/s1, Table S1: Enological parameters of fresh Corvina and Corvinone musts at crush; Table S2: Quantification methods, retention indices, quantification ions of studied compounds; Figure S1: Fermentation kinetics of each biological replicate (a and b) of all wines; Table S3: Kruskal–Wallis ($p < 0.05$) of volatile compounds according to employed yeasts and grape origin in wines; Table S4: Significantly different compounds according to Kruskal–Wallis analysis ($\alpha = 0.05$) between Spontaneous and inoculated (yeast 1, yeast 2, yeast 3, yeast 4) fermentations in wines; Table S5: Concentration (μg/L) of significant different volatile compounds among different sensory clusters according to Kruskal–Wallis ($\alpha = 0.05$).

Author Contributions: Conceptualization, M.U. and G.L.; methodology, M.U., D.S. and G.L.; validation, M.U.; formal analysis, G.L.; resources, M.U.; data curation, G.L., D.S. and M.U.; writing—original draft preparation, G.L.; writing—review and editing, M.U. and D.S.; visualization, M.U.; supervision, M.U. and D.S.; project administration, M.U.; funding acquisition, M.U. All authors have read and agreed to the published version of the manuscript.

Funding: This research received no external funding.

Informed Consent Statement: Informed consent was obtained from all subjects involved in the study.

Data Availability Statement: Data is contained within the article and supplementary material.

Acknowledgments: The authors thank the winery Fratelli Tedeschi for financial support and for providing the grapes employed.

Conflicts of Interest: The authors declare no conflict of interest.

References

1. *Disciplinare di Produzione dei vini a Denominazione di Origine. Controllata e Garantita "Amarone della Valpolicella". (Gazzetta Ufficiale 190, 14.08.2019)*; Ministero delle Politiche Agricole Alimentari e Forestali; Dipartimento delle Politiche Competitive, della Qualità Agroalimentare, Ippiche e della Pesca: Rome, Italy, 2019.
2. Bellincontro, A.; Matarese, F.; D'Onofrio, C.; Accordini, D.; Tosi, E.; Mencarelli, F. Management of postharvest grape withering to optimise the aroma of the final wine: A case study on Amarone. *Food Chem.* **2016**, *213*, 378–387. [CrossRef]
3. Fedrizzi, B.; Tosi, E.; Simonato, B.; Finato, F.; Cipriani, M.; Caramia, G.; Zapparoli, G. Changes in wine aroma composition according to botrytized berry percentage: A preliminary study on Amarone wine. *Food Technol. Biotechnol.* **2010**, *49*, 429–536.
4. Slaghenaufi, D.; Guardini, S.; Tedeschi, R.; Ugliano, M. Volatile terpenoids, norisoprenoids and benzenoids as markers of fine scale vineyard segmentation for Corvina grapes and wines. *Food Res. Int.* **2019**, *125*, 108507. [CrossRef]
5. Luzzini, G.; Slaghenaufi, D.; Pasetto, F.; Ugliano, M. Influence of grape composition and origin, yeast strain and spontaneous fermentation on aroma profile of Corvina and Corvinone wines. *Lebensm.-Wiss. Technol.* **2021**, *143*, 111120. [CrossRef]
6. Slaghenaufi, D.; Boscaini, A.; Prandi, A.; Dal Cin, A.; Zandonà, V.; Luzzini, G.; Ugliano, M. Influence of Different Modalities of Grape Withering on Volatile Compounds of Young and Aged Corvina Wines. *Molecules* **2020**, *25*, 2141. [CrossRef] [PubMed]
7. Slaghenaufi, D.; Peruch, E.; De Cosmi, M.; Nouvelet, L.; Ugliano, M. Volatile and phenolic composition of monovarietal red wines of Valpolicella appellations. *Oeno One* **2021**, *55*, 279–294. [CrossRef]
8. Bayonove, C. L'arome varietal: Le potentiel aromatique du raisin. In *Oenologie: Fondements Scientifiques et Technologiques*; Flanzy, C., Ed.; Lavoisier Tec & Doc: Paris, France, 1998; pp. 165–181.
9. Kotseridis, Y.; Baumes, R. Identification of Impact Odorants in Bordeaux Red Grape Juice, in the Commercial Yeast Used for Its Fermentation, and in the Produced Wine. *J. Agric. Food Chem.* **2000**, *48*, 400–406. [CrossRef] [PubMed]
10. Etievant, P.X. Wine. In *Volatile Compounds*; Maarse, H., Ed.; M. Dekker: New York, NY, USA, 1991; pp. 483–546.
11. Ferreira, V. Volatile aroma compounds and wine sensory attributes. In *Managing Wine Quality*; Reynolds, A.G., Ed.; Woodhead Publishing Series: Amsterdam, The Netherlands, 2010; pp. 3–28.
12. Seguin, G. Ecosystems of the great red wines produced in the maritime climate of Bordeaux. In *Proceedings of the Symposium on Maritime Climate Winegrowing*; Department of Horticultural Sciences, Cornell University: Ithica, NY, USA, 1988.
13. Parker, M.; Capone, D.; Francis, I. Aroma Precursors in Grapes and Wine: Flavor Release during Wine Production and Consumption. *J. Agric. Food Chem.* **2017**, *66*, 2281–2286. [CrossRef]
14. Ugliano, M.; Henschke, P. Yeasts and Wine Flavour. In *Wine Chemistry and Biochemistry*; Moreno-Arribas, M.V., Polo., C., Eds.; Springer: New York, NY, USA, 2009; pp. 313–392.
15. Slaghenaufi, D.; Indorato, C.; Troiano, E.; Luzzini, G.; Felis, G.; Ugliano, M. Fate of Grape-Derived Terpenoids in Model Systems Containing Active Yeast Cells. *J. Agric. Food Chem.* **2020**, *68*, 13294–13301. [CrossRef] [PubMed]
16. Robinson, A.; Boss, P.; Solomon, P.; Trengove, R.; Heymann, H.; Ebeler, S. Origins of Grape and Wine Aroma. Part 1. Chemical Components and Viticultural Impacts. *J. Enol. Vitic.* **2014**, *65*, 1–24. [CrossRef]
17. Antalick, G.; Šuklje, K.; Blackman, J.; Meeks, C.; Deloire, A.; Schmidtke, L. Influence of Grape Composition on Red Wine Ester Profile: Comparison between Cabernet Sauvignon and Shiraz Cultivars from Australian Warm Climate. *J. Agric. Food Chem.* **2015**, *63*, 4664–4672. [CrossRef]
18. Lambrechts, M.; Pretorius, I. Yeast and its Importance to Wine Aroma—A Review. *S. Afr. J. Enol. Vitic.* **2000**, *21*, 97–129. [CrossRef]
19. Moio, L.; Ugliano, M.; Genovese, A.; Gambuti, A.; Pessina, R.; Piombino, P. Effect of Antioxidant Protection of Must on Volatile Compounds and Aroma Shelf Life of Falanghina (*Vitis vinifera* L.) Wine. *J. Agric. Food Chem.* **2004**, *52*, 891–897. [CrossRef]
20. Genovese, A.; Caporaso, N.; Moio, L. Influence of Yeast Strain on Odor-Active Compounds in Fiano Wine. *Appl. Sci.* **2021**, *11*, 7767. [CrossRef]
21. Varela, C.; Barker, A.; Tran, T.; Borneman, A.; Curtin, C. Sensory profile and volatile aroma composition of reduced alcohol Merlot wines fermented with Metschnikowia pulcherrima and *Saccharomyces* uvarum. *Int. J. Food Microbiol.* **2017**, *252*, 1–9. [CrossRef] [PubMed]
22. Varela, C.; Bartel, C.; Espinase Nandorfy, D.; Bilogrevic, E.; Tran, T.; Heinrich, A.; Balzan, T.; Bindon, K.; Borneman, A. Volatile aroma composition and sensory profile of Shiraz and Cabernet Sauvignon wines produced with novel Metschnikowia pulcherrima yeast starter cultures. *Aust. J. Grape Wine Res.* **2021**, *27*, 406–418. [CrossRef]
23. Swiegers, J.P. Modulation of volatile sulfur compounds by wine yeast. *Appl. Microbiol. Biotechnol.* **2007**, *74*, 954–960.er. [CrossRef]
24. Ugliano, M.; Travis, B.; Francis, I.; Henschke, P. Volatile Composition and Sensory Properties of Shiraz Wines as Affected by Nitrogen Supplementation and Yeast Species: Rationalizing Nitrogen Modulation of Wine Aroma. *J. Agric. Food Chem.* **2010**, *58*, 12417–12425. [CrossRef] [PubMed]
25. Binati, R.; Lemos Junior, W.; Luzzini, G.; Slaghenaufi, D.; Ugliano, M.; Torriani, S. Contribution of non-*Saccharomyces* yeasts to wine volatile and sensory diversity: A study on Lachancea thermotolerans, Metschnikowia spp. and Starmerella bacillaris strains isolated in Italy. *Int. J. Food Microbiol.* **2020**, *318*, 108470. [CrossRef]
26. Tufariello, M.; Fragasso, M.; Pico, J.; Panighel, A.; Castellarin, S.; Flamini, R.; Grieco, F. Influence of Non-*Saccharomyces* on Wine Chemistry: A Focus on Aroma-Related Compounds. *Molecules* **2021**, *26*, 644. [CrossRef]
27. Zenoni, S.; Fasoli, M.; Guzzo, F.; Santo, S.; Amato, A.; Anesi, A.; Commisso, M.; Herderich, M.; Ceoldo, S.; Avesani, L.; et al. Disclosing the molecular basis of the postharvest life of berry in different grapevinegenotypes. *Plant Physiol.* **2016**, *172*, 1821–1843. [CrossRef] [PubMed]

28. Zamboni, A.; Minoia, L.; Ferrarini, A.; Tornielli, G.B.; Zago, E.; Delledonne, M.; Pezzotti, M. Molecular analysis of post-harvest withering in grape by AFLP transcriptional profiling. *J. Exp. Bot.* **2008**, *59*, 4145–4159. [CrossRef] [PubMed]
29. Bely, M.; Rinaldi, A.; Dubourdieu, D. Influence of assimilable nitrogen on volatile acidity production by *Saccharomyces cerevisiae* during high sugar fermentation. *J. Biosci. Bioeng.* **2003**, *96*, 507–512. [CrossRef]
30. Mateo, J.; Jiménez, M.; Pastor, A.; Huerta, T. Yeast starter cultures affecting wine fermentation and volatiles. *Food Res. Int.* **2001**, *34*, 307–314. [CrossRef]
31. Boss, P.K.; Böttcher, C.; Davies, C. Various Influences of Harvest Date and Fruit Sugar Content on Different Wine Flavor and Aroma Compounds. *Am. J. Enol. Vitic.* **2014**, *65*, 341–353. [CrossRef]
32. Slaghenaufi, D.; Ugliano, M. Norisoprenoids, Sesquiterpenes and Terpenoids Content of Valpolicella Wines During Aging: Investigating Aroma Potential in Relationship to Evolution of Tobacco and Balsamic Aroma in Aged Wine. *Front. Chem.* **2018**, *6*, 66. [CrossRef]
33. Alegre, Y.; Sáenz-Navajas, M.; Hernández-Orte, P.; Ferreira, V. Rapid strategies for the determination of sensory and chemical differences between a wealth of similar wines. *Eur. Food Res. Technol.* **2017**, *243*, 1295–1309. [CrossRef]
34. Parr, W.; Heatherbell, D.; White, K.G. Demystifying Wine Expertise: Olfactory Threshold, Perceptual Skill and Semantic Memory in Expert and Novice Wine Judges. *Chem. Senses* **2002**, *27*, 747–755. [CrossRef]
35. Barbanti, D.; Mora, B.; Ferrarini, R.; Tornielli, G.; Cipriani, M. Effect of various thermo-hygrometric conditions on the withering kinetics ofg rapes used for the production of "Amarone" and "Recioto" wines. *J. Food Eng.* **2008**, *85*, 350–358. [CrossRef]
36. Amerine MARoessler, E.B. *Wines: Their Sensory Evaluation*; W. H. Freeman & Co.: New York, NY, USA, 1980.
37. Celik, Z.D.; Karaoglan, S.Y.; Darıcı, M.; Kelebek, H.; İşçi, B.; Kaçar, E.; Altındişli, A.; Cabaroglu, T. Effects of terroir on the terpene compounds of Muscat of Bornova Native white grape variety grown in Turkey. *BIO Web Conf.* **2015**, *5*, 01004. [CrossRef]
38. Sabon, I.; De Revel, G.; Kotseridis, Y.; Bertrand, A. Determination of volatile compounds in Grenache wines in relation with different terroirs in the Rhone Valley. *J. Agric. Food Chem.* **2002**, *50*, 6341–6345. [CrossRef] [PubMed]
39. Villanova, M.; Zamuz, S.; Vilariño, F.; Sieiro, C. Effect of terroir on the volatiles of *Vitis vinifera* cv. Albariño. *J. Agric. Food Chem.* **2007**, *87*, 1252–1256. [CrossRef]
40. Úbeda, C.; del Barrio-Galán, R.; Peña-Neira, Á.; Medel-Marabolí, M.; Durán-Guerrero, E. Location Effects on the Aromatic Composition of Monovarietal cv. Carignan Wines. *Am. J. Enol. Vitic.* **2017**, *68*, 390–399. [CrossRef]
41. Black, C.; Parker, M.; Siebert, T.; Capone, D.; Francis, I. Terpenoids and their role in wine flavour: Recent advances. *Aust. J. Grape Wine Res.* **2015**, *21*, 582–600. [CrossRef]
42. Marais, J.; van Wyk, C.; Rapp, A. Effect of Sunlight and Shade on Norisoprenoid Levels in Maturing Weisser Riesling and Chenin blanc Grapes and Weisser Riesling wines. *S. Afr. J. Enol. Vitic.* **1992**, *13*, 23–32.
43. Marais, J.J.; van Wyk, C. Effect of Grape Maturity and Juice Treatments on Terpene. *S. Afr. J. Enol. Vitic.* **1986**, *7*, 26–35.
44. Ugliano, M.; Bartowky, E.; McCarthy, J.; Moio, L.; Henschke, P. Hydrolysis and Transformation of Grape Glycosidically Bound Volatile Compounds during Fermentation with Three *Saccharomyces* Yeast Strains. *J. Agric. Food Chem.* **2006**, *54*, 6322–6331. [CrossRef] [PubMed]
45. Lloyd, N.; Capone, D.; Ugliano, M.; Taylor, D.; Skouroumounis, G.; Sefton, M.; Elsey, G. Formation of Damascenone under both Commercial and Model Fermentation Conditions. *J. Agric. Food Chem.* **2011**, *59*, 1338–1343. [CrossRef]
46. Kruis, A.J.; Levisson, M.; Mars, A.E.; der Ploeg, M.; Garcés Daza, F.; Ellena VKengen, S.W.M.; der Oost, J.; Weusthuis, R.A. Ethyl acetate production by the elusive alcohol acetyltransferase from yeast. *Metab. Eng.* **2017**, *41*, 92–101. [CrossRef]
47. Löser, C.; Urit, T. Perspectives for the biotechnological production of ethyl acetate by yeasts. *Appl. Microbiol. Biotechnol.* **2014**, *98*, 5397–5415. [CrossRef] [PubMed]
48. Fleet, G. *Wine Microbiology and Biotechnology*; CRC Press: Boca Raton, FL, USA, 1993.
49. Valentin, D.C.; Nestrud, M.; Abdi, H. Projective Mapping and Sorting Tasks. In *Descriptive Analysis in Sensory Evaluation*; Hort, J., Kemp, S., Hollowood, T., Eds.; Wiley-Blackwell: London, UK, 2016.
50. Francesca, N.; Gaglio, R.; Alfonzo, A.; Settanni, L.; Corona, O.M.; Romano, R.; Piccolo, A.; Moschetti, G. The Wine: Typicality or mere diversity? The effect of spontaneous. *Agric. Sci. Procedia* **2016**, *8*, 769–773.
51. Romano, P.; Fiore, C.; Paraggio, M.; Caruso, M.; Capece, A. Function of yeast species and strains in wine flavour. *Int. J. Food Microbiol.* **2003**, *86*, 169–180. [CrossRef]
52. Zorhe, D.; Erten, H. The influence of *Kloeckera apiculata* and *Candida pulcherrima* yeasts on wine fermentation. *Process Biochem.* **2002**, *38*, 318–324. [CrossRef]

Article

Effect of Microwave Maceration and SO_2 Free Vinification on Volatile Composition of Red Wines

Raquel Muñoz García [1], Rodrigo Oliver Simancas [1], María Consuelo Díaz-Maroto [1,*], María Elena Alañón Pardo [2] and María Soledad Pérez-Coello [1]

[1] Area of Food Technology, Faculty of Chemical Sciences and Technologies, Regional Institute for Applied Scientific Research (IRICA), University of Castilla-La Mancha, Avda. Camilo José Cela 10, 13071 Ciudad Real, Spain; raquel.munoz@uclm.es (R.M.G.); rodrigo.oliver@uclm.es (R.O.S.); soledad.perez@uclm.es (M.S.P.-C.)

[2] Area of Food Technology, Higher Technical School of Agronomic Engineering, University of Castilla-La Mancha, Ronda de Calatrava 7, 13071 Ciudad Real, Spain; mariaelena.alanon@uclm.es

* Correspondence: mariaconsuelo.diaz@uclm.es

Citation: Muñoz García, R.; Oliver Simancas, R.; Díaz-Maroto, M.C.; Alañón Pardo, M.E.; Pérez-Coello, M.S. Effect of Microwave Maceration and SO_2 Free Vinification on Volatile Composition of Red Wines. *Foods* **2021**, *10*, 1164. https://doi.org/10.3390/foods10061164

Academic Editors: Fulvio Mattivi and Matteo Bordiga

Received: 20 April 2021
Accepted: 19 May 2021
Published: 22 May 2021

Abstract: This study evaluates the effect of microwave treatment in grape maceration at laboratory scale on the content of free and glycosidically bound varietal compounds of must and wines and on the overall aroma of wines produced with and without SO_2. The volatile compounds were extracted by solid phase extraction and analyzed by gas chromatography–mass spectrometry, carrying out a sensory evaluation of wines by quantitative descriptive analysis. Microwave treatment significantly increased the free and bound fraction of most varietal compounds in the must. Wines from microwave maceration showed faster fermentation kinetics and shorter lag phase, resulting in an increase in some volatile compounds of sensory relevance. The absence of SO_2 caused a decrease in concentration of some volatile compounds, mainly fatty acids and esters. The sensory assessment of wines from microwave treatment was higher than the control wine, especially in wines without SO_2, which had higher scores in the "red berry" and "floral" odor attributes and a more intense aroma. This indicates that the pre-fermentative treatment of grapes with microwaves could be used to increase the wine aroma and to reduce the occurrence of SO_2.

Keywords: microwave maceration; red wine; volatile compounds; aroma

1. Introduction

For decades, the wine industry has used various technological innovations to improve the quality of its wines, but also to achieve environmentally sustainable processes that reduce the use of potentially hazardous products for health. In this sense, emerging technologies based on the use of ultrasound, microwave, or electric fields have been revealed as clean technologies that thanks to their mechanical effect on cells achieve greater performance in the extraction of compounds during maceration processes (phenolic, volatile, bioactive compounds...) [1]. These techniques can also help reduce the presence of spoilage or pathogen microorganisms, avoiding the excessive use of antiseptic agents such as SO_2, and obtaining higher quality and healthier wines [2,3].

The aroma of wine is one of the attributes most appreciated by consumers, this aroma is formed by numerous compounds that come in part from grapes (varietal compounds) and are influenced by grape variety and agro-climatic factors. Many of the varietal compounds are found in the grape skin, either in free form as directly contributing to wine aroma or as non-volatile precursors, which release odorous molecules through enzymatic or chemical hydrolysis reactions [4,5]. Yeast metabolism also generates many volatile compounds during the alcoholic fermentation, which will be affected by the type of yeast, as well as fermentation conditions, while other aromas develop during the aging process.

Several techniques have been used to promote the extraction of compounds from grape skin during the maceration process such as the use of pectolytic enzymes, cryomaceration, or thermovinification [1,6,7]. More recently, novel techniques such as ultrasound or pulsed electric fields have been applied to red winemaking to increase the extraction of phenolic compounds and to reduce maceration and aging times [8–11]. However, its effect on the fraction of volatile compounds has been less studied. Cryomaceration has been described as promoting the extraction of varietal compounds in grapes [12]. Meanwhile, thermal processes such as thermovinification can affect the degradation of volatile compounds in the wine or promote its extraction, depending on the conditions used [13,14]. The effect of sonication during grape maceration can lead to greater extraction of the free and bound fraction of the aroma and an increase in the aroma of the final wine [15,16], although some authors describe the formation of off-flavor or do not notice changes in the aroma of wine due to sonication [17].

Microwave technology (MW) has been applied in the food industry to reduce processing times and food preservation [1]. Microwaves are non-ionizing electromagnetic waves with frequencies between 300 MHz and 300 GHz, although the frequency range for industrial use is smaller (0.9–2.45 GHz). Microwaves have a double effect on matrices: molecular movement by the migration of ions and rotation of molecules with permanent dipoles formation. The resistance offered to ion migration and the realignment of the dipoles generates friction forces that originate heating, without altering their molecular structures unless the temperature is too high [18].

This thermal energy can disrupt solute bonds with the matrix, favoring its release and helping the extraction process. If the temperature is high, a explosion of grape cells can occur by releasing its contents into the liquid phase. Microwave-assisted extraction (MAE) has been used for the extraction of analytes (volatile compounds, bioactive phenolics) in plants and wines, achieving better yields than conventional processes [19–21].

MW technology has been applied to reduce winemaking time by achieving higher overall wine quality [22,23]. In wines from Pinot Noir and Merlot varieties, grape maceration applying microwave succeeded in increasing the amount of total phenolic compounds, anthocyanins, and tannins of wine by shortening maceration times [24–26]. These authors observed an increase in the floral and fruity aroma in MW-treated wines, although volatile compounds were not analyzed. However, Tartian et al. [27] observed that microwave maceration was less effective in the extraction of volatile compounds of sensory relevance to cryomaceration or traditional maceration.

Microwave pulses have also been proposed as a method to accelerate the aging of wine as well as to favor the transfer of wood compounds in treatments with oak chips [10].

On the other hand, radiation with microwaves helps to preserve food, reducing the presence of spoilage or pathogen microorganisms [28]. Based on this fact, microwave treatment has been proposed in the oenological industry for the sanitation of reused oak barrels, achieving a reduction in the population of Brettanomyces and lactic bacteria [29]. Carew et al. [24,25] proposed the use of high-power microwaves in grape crushing inoculating with a starter culture without SO_2 addition, seeing that fermentation was developing correctly, obtaining faster fermentation kinetics and higher yield.

The objective of this work was to study the effect of the microwave treatment at laboratory scale with temperature control during the pre-fermentative maceration of Cabernet Sauvignon grapes, on the extraction of the free and glycosidically bound fraction of volatile compounds and on the overall aroma of wines, produced with and without SO_2 addition, with a view to its possible reduction in vinification.

2. Materials and Methods

2.1. Microwave Treatment and Microvinification

Cabernet Saugvinon red grapes were harvested at the optimal state of maturity during the 2019 harvest at the Institute of Vine and Wine of Castilla-La Mancha (IVICAM, Tomelloso, Ciudad Real, Spain). They were immediately transported to the laboratory for their processing. The chemical composition of the must was 24.9°Brix, 3.81 g/L of total acidity, and 3.26 of pH.

Grapes (150 kg) were destemmed and crushed and then divided into three batches. Two batches were sulphited at this time with 50 mg/L SO_2, and the third batch was microwave-treated without SO_2. One batch with SO_2 ("C") was not submitted to any microwave treatment and was used as control wine. This sample was immediately moved to a chamber at 22 °C (±2 °C). The other two batches ("MW") were microwave macerated during 12 min at 700 W using a domestic LG MJ3965ACS microwave oven (LG electronics, Madrid, Spain). The samples were processed in batches of 2 kg. The temperature on the batch prior to the treatment was 14 °C, which was raised up to 40 °C when it finished. To avoid an increment up to 40 °C, the microwave treatment was realized at 3 intervals of 4 min each up, to a total of 12 min. At the end of each time interval, the batches were stirred, and their temperatures were evaluated using a solid stem thermometer. All treatments were executed by triplicate. Then, microwave treatment batches were moved to the 22 °C (±2 °C) chamber for sampling, yeast inoculation, and fermentation.

Alcoholic fermentation–maceration was realized in 10 L glass containers that were filled with 6 kg of sample (3.6 kg of must and 2.4 kg of skins (in order to not filled more than 60 %) using *Saccharomyces cerevisiae* (CECT no. 10835) as starter culture, at 22 °C (±2 °C). They were punched down twice every day. The evolution of the fermentation was controlled by weight loss, and it was considered finished when the container weight remained stable. The final of the fermentation was confirmed by density and glucose and fructose analysis of wines. Finally, wines were decanted, filtered, and bottled. Only wines with SO_2 were adjusted to 25 mg/L of free SO_2 before bottling when it was necessary.

2.2. Solid Phase Extraction

Solid phase extraction (SPE) was used for the isolation of the free and glycosidically bound volatile compounds of musts and wines. Before SPE, musts were centrifuged at 4 °C (10,000 rpm, 10 min) (Avanti Centrifuge J26-XP, Beckman Coulter, CA, USA) and filtrated through a 1.2 μm glass fiber membrane (Fisherbrand, Thermo Fisher Scientific, Inc., Waltham, MA, USA), while wines did not need prior preparation. All extractions were carried out in duplicate according to the method described by Oliver Simancas et al. [16].

2.2.1. Free Volatile Compounds

First, the SPE cartridges (500 mg styrene-divinylbenzene, Lichrolut EN Merck, KGaA, Darmstadt, Germany) were conditioned with 10 mL of dichloromethane, 5 mL of methanol, and 10 mL ethanol:water (10:90, *v*/*v*). Then, 100 mL of sample together with 40 μL of the internal standard (4-nonanol, 1 g/L) were passed through the cartridge. Non-volatile hydrophilic compounds were washed out of the cartridges with 50 mL of bidistilled water (Milli Q Plus, Merck KGaA, Darmstadt, Germany), and the free volatile compounds were eluted with 10 mL of dichloromethane. The extracts were concentrated to an approximate volume of 200 μL under nitrogen and stored at −20 °C until their analysis.

2.2.2. Glycosidically Bound Volatile Compounds

Once the free volatile compounds were removed, glycosidically bound fraction was eluted from the same cartridge with 25 mL of ethyl acetate:methanol (90:10, *v*/*v*). Extracts were evaporated under vacuum (55 °C, 290 mBar) to almost dryness (≈1 mL), achieving complete dryness under a gentle stream of nitrogen. To recover the dry extract distributed in the flask walls, it was re-dissolved with 1 mL of methanol and evaporated again with nitrogen until dryness. The enzymatic hydrolysis was performed by adding 0.20 μL/mL of a commercial enzyme "Trenolin Bouquet PLUS" (ERBSLÖH, Geisenheim, Germany) in 2 mL of phosphate-citrate buffer (0.1–0.2 M; pH = 5),and incubating at 40 °C for 16 h. Then, 25 mL of synthetic wine (12% ethanol, pH = 3.5) were added to the flask, and the released volatile compounds were recovered by SPE, using 200 mg styrene-divinylbenzene cartridges (Lichrolut EN Merck, KGaA, Darmstadt, Germany) previously conditioned (5 mL of dichloromethane, 2.5 mL of methanol, and 5 mL of 10 % ethanol:water). Hydrophilic compounds were removed from the cartridges with 25 mL of bidistilled water (Milli Q Plus, Merck KGaA, Darmstadt, Germany), and released volatile compounds were eluted with 5 mL of dichloromethane. The extracts were concentrated to an approximate volume of 200 μL under nitrogen and stored at −20 °C until their analysis.

2.3. Gas Chromatography-Mass Spectrometry

Extracts (1 μL) were analyzed in an Agilent 6890 GC System coupled to an Agilent 5973 inert Mass Selective Detector and equipped with a DB-WAX column (60 m × 0.25 mm × 0.25 μm) (Agilent Technologies, Inc.), using helium as the carrier gas (1 mL/min). The oven temperature program started at 70 °C for 5 min; then, it was raised up first at 1 °C/min to 90 °C (10 min) and then at 2 °C/min to 210 °C, which was maintained for 40 min. The injector temperature was 250 °C. The MS conditions were: electron impact mode (70 eV), ion source temperature of 230 °C, and scanning from 45 to 550 a.m.u.

Identification of the volatile compounds was carried out by comparison with standards from Sigma-Aldrich (Tres Cantos, Madrid, Spain). Compounds for which it was not possible to find volatile references were tentatively identified by comparison of their mass spectra with those from different mass spectrum libraries (Wiley G 1035 A, NBS75K and NIST14). Quantitative analysis (μg/L) was performed considering the response factors of the different volatile compounds, except for those whose commercial standards were not available, which were quantified by means of the response factors of compounds with similar chemical structures: geranial was quantified as *trans*-geraniol; 3-oxo-α-ionol, 3-hydroxy-7,8-dihydro-β-ionol, and 6,7-dehydro-7,8-dihydro-3-oxo-ionol were quantified as α-ionol; 4-methyl-1-pentanol was quantified as 3-methyl-1-pentanol; isovaleric acid was quantified as propanoic acid; isobutyric acid was quantified as butanoic acid; ethyl 3-hydroxyhexanoate was quantified as ethyl hexanoate; and methyl 2-hydroxy-4-methylpentanoate and ethyl 4-hydroxybutyrate were quantified as ethyl 3-hydroxybutyrate.

2.4. Descriptive Sensory Analysis

Wines were tasted, in duplicate, three months after bottling, by eight judges with high experience in wine sensory analysis, two men and six women, aged between 23 and 62, belonging to the staff of the Food Science and Technology Area of the University of Castilla-La Mancha. Evaluation took place in a standard sensory analysis chamber (ISO 8589:2007) equipped with separate booths. While wines were presented in standard wine-tasting glasses (ISO 3591:1997). Samples were sniffed and tasted, and the judges generated sensory terms individually, agreeing on the following olfactory descriptors: floral, green, red berry, prune, and odor intensity. A 10 cm unstructured scale was used to measure each attribute, indicating in the left extreme "attribute not perceptible" and in the right extreme "attribute strongly perceptible".

2.5. Statistical Analysis

Student–Newman–Keuls's test was used to find significant differences between the volatile compounds of the samples, while ANOVA analysis was applied to sensory data. Both analyses were carried out with the IBM SPSS statistical package, version 24.0 (IBM, Madrid, Spain).

3. Results and Discussion

3.1. Effect of Microwave Treatment on Volatile Compounds of Musts

Figure 1a,b shows the main groups of varietal volatile compounds quantified in the free and bound fraction of the musts, respectively. Aldehydes and C6 alcohols were the majority groups. These compounds are formed by the degradation of lipids from the grape skin in the presence of oxygen, mainly in pre-fermentation processes.

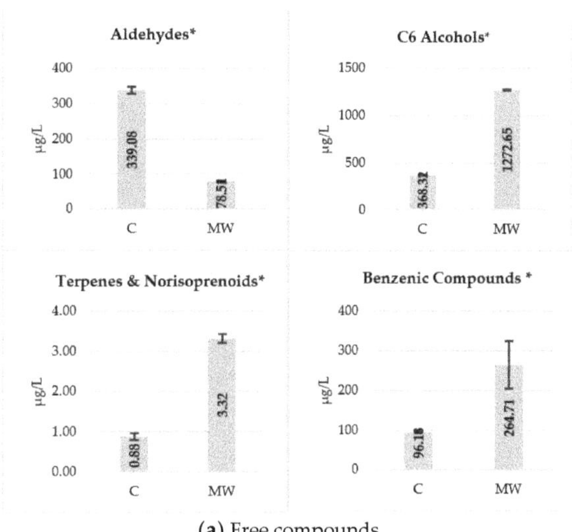

(a) Free compounds

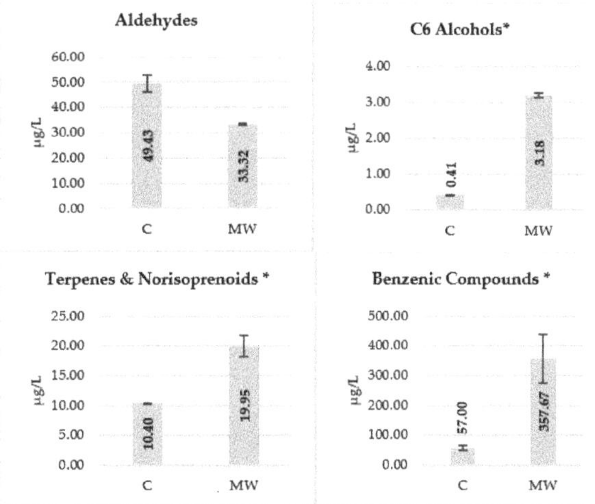

(b) Bound compounds

Figure 1. Mean concentrations (μg/L) of the main groups of free (**a**) and bound compounds (**b**) in control must (C) and must from microwave-treated grapes (MW). *: denote significant differences according to Student's t-test ($p \leq 0.05$).

Hexanal was the majority aldehyde in the control must followed by trans-2-hexenal (Table S1 Supplementary Material). The aldehydes are characterized by their pleasant aromas (sweet, orange) and low olfactory detection thresholds but, similar to other low molecular weight aldehydes, they are transformed during pre-fermentative and fermentative processes into their corresponding alcohols [4].

It should be noted that free aldehydes had much lower concentrations in microwave-treated musts (Figure 1a), whereas the opposite occurred with the corresponding alcohols. Among these, 1-hexanol and *cis*-2-hexen-1-ol were the majority compounds in musts treated with microwaves. This suggests that microwave treatment significantly increasing their extraction from grape skin, especially in the case of alcohols.

The bound fraction of the aldehydes and C6 alcohols was much lower than the free fraction, which has been previously described [4], and the effect of microwave maceration was not significant.

Terpenes and norisoprenoids are mainly in their bound form and may also be affected by intensive maceration techniques [30]. In fact, although they were found in small amounts, their concentration in both free and bound form increased in the musts treated with MW.

Benzenic compounds had similar amounts in their free and bound form, with a very positive effect of microwave treatment observed in the extraction of both fractions, which greatly increased in their bound form. This effect was also observed in grape ultrasonic maceration [16]. The major compounds were benzyl alcohol and 2-phenylethanol, as well as others such as 4-vinylguaiacol or vanillin (Tables S1 and S2, Supplementary Material), whose quantities in wines will be increased by the metabolism of yeasts and may reach concentrations of sensory significance.

No previous work has been found on the effect of using MW in grape maceration on volatile compounds of musts. However, further extraction of phenolic compounds and amino acids in high-powered MW maceration has been described, by proposing this technique for obtaining Pinot Noir wines with greater color reducing maceration times [23,24].

Other authors have observed an increase in the free and bound fraction of must volatiles applying various techniques in the grape maceration step as ultrasounds and pulsed electric fields [8,16,31].

Likewise, the effect of microwaves on skin cells has been able to facilitate the release of volatile compounds free and glycosidically bound into the liquid phase. In any case, the use of maceration with MW would increase the aromatic potential of grape musts.

3.2. *Effect of Microwave Treatment on Volatile Compounds of Wines Fermented with and without SO_2.*

The grape must from pre-fermentative maceration with MW and the control must were vinified in the traditional way with SO_2 addition. However, due to the inhibitory effect of microwaves on the native microbial population described by some authors [24,25], parallel fermentation of the microwave-treated must without SO_2 was carried out. Tables 1 and 2 show the results of the free and glycosidically bound fraction of the varietal compounds quantified in wines.

Table 1. Free volatile compound concentrations (μg/L) in control wines and those obtained from microwave-treated grapes elaborated with and without SO_2 (n = 3).

Volatile Compounds	Control Wine	MW Wine	MW Wine without SO_2
	Mean ± SD	Mean ± SD	Mean ± SD
1-hexanol	227.32 ± 48.70 [a]	812.81 ± 33.61 [c]	553.42 ± 76.76 [b]
cis-3-hexen-1-ol	19.92 ± 2.00 [a]	41.78 ± 2.35 [b]	17.09 ± 1.67 [a]
trans-3-hexen-1-ol	4.25 ± 0.58 [a]	10.54 ± 0.72 [b]	4.92 ± 0.49 [a]
cis-2-hexen-1-ol	8.94 ± 1.06 [c]	4.18 ± 1.06 [b]	1.48 ± 0.28 [a]
trans-2-hexen-1-ol	2.89 ± 0.69 [a]	4.97 ± 0.38 [b]	4.02 ± 0.88 [a,b]
Σ C6 alcohols	263.32 ± 50.24 [a]	874.28 ± 37.19 [c]	580.93 ± 77.86 [b]

Table 1. *Cont.*

Volatile Compounds	Control Wine	MW Wine	MW Wine without SO_2
	Mean ± SD	Mean ± SD	Mean ± SD
linalool	5.65 ± 0.99 [a]	5.61 ± 0.15 [a]	12.00 ± 0.58 [b]
α-terpineol	3.10 ± 0.37 [a]	2.94 ± 0.84 [a]	3.74 ± 0.26 [a]
β-damascenone	72.97 ± 2.20 [b]	25.79 ± 2.80 [a]	23.12 ± 1.74 [a]
trans-geraniol	10.51 ± 2.69 [a]	16.07 ± 2.02 [b]	9.77 ± 2.59 [a]
geranial	40.27 ± 6.03 [b]	132.88 ± 8.47 [c]	15.88 ± 1.14 [a]
nerolidol	20.91 ± 3.43 [a]	29.97 ± 0.26 [a]	42.17 ± 7.38 [b]
6,7-dehydro-7,8-dihydro-3-oxo-ionol	33.44 ± 4.05 [b]	23.53 ± 0.22 [a]	28.24 ± 2.09 [a,b]
3-oxo-α-ionol	102.25 ± 2.09 [a]	100.39 ± 26.65 [a]	60.97 ± 12.63 [a]
Σ Terpenes and norisoprenoids	289.11 ± 2.58 [b]	337.18 ± 19.73 [c]	195.89 ± 8.72 [a]
benzaldehyde	6.53 ± 1.20 [a]	5.40 ± 0.21 [a]	6.31 ± 1.52 [a]
guaiacol	45.59 ± 13.28 [b]	25.75 ± 3.10 [a]	54.59 ± 5.01 [b]
benzyl alcohol	184.07 ± 46.40 [b]	105.67 ± 3.38 [a]	214.91 ± 8.77 [b]
vinylguaicol	72.25 ± 13.59 [b]	27.08 ± 3.73 [a]	64.14 ± 5.35 [b]
syringol	170.76 ± 32.34 [a]	222.23 ± 23.22 [a]	187.71 ± 12.31 [a]
vanillin	7.71 ± 1.66 [a]	6.31 ± 1.15 [a]	6.27 ± 0.31 [a]
methyl vanillate	77.23 ± 17.87 [b]	37.60 ± 2.36 [a]	44.89 ± 5.73 [a]
ethyl vanillate	3.15 ± 0.11 [a]	24.54 ± 0.81 [b]	95.49 ± 4.11 [c]
methyl vanillyl ether	76.73 ± 5.23 [a]	79.13 ± 3.44 [a]	84.09 ± 15.05 [a]
Σ Benzenic compounds	644.00 ± 99.16 [a,b]	533.72 ± 27.22 [a]	758.40 ± 29.56 [b]

Values with different superscripts in the same row denoted significant differences according to the Student–Newman–Keuls test at $p < 0.05$. Samples were defined according to the treatment applied: "Control wine": no treatment applied; "MW wine": microwave treatment at 700 W, 12 min of grape crushed; "MW wine without SO_2": microwave treatment at 700 W, 12 min of grape crushed elaborated without SO_2.

Table 2. Glycosidically bound volatile compound concentrations (μg/L) in control wines and those obtained from microwave-treated grapes elaborated with and without SO_2 (n = 3).

Volatile Compounds	Control Wine	MW Wine	MW Wine without SO_2
	Mean ± SD	Mean ± SD	Mean ± SD
1-hexanol	3.73 ± 0.72 [a]	5.47 ± 1.33 [a]	4.43 ± 1.04 [a]
trans-3-hexen-1-ol	0.73 ± 0.10 [a]	6.04 ± 0.08 [b]	0.79 ± 0.18 [a]
cis-2-hexen-1-ol	2.10 ± 0.17 [c]	0.83 ± 0.05 [a]	1.60 ± 0.07 [b]
trans-2-hexen-1-ol	0.66 ± 0.13 [b]	0.39 ± 0.01 [a]	0.56 ± 0.04 [b]
Σ C6 alcohols	7.22 ± 0.82[a]	12.73 ± 1.25 [b]	7.38 ± 1.02 [a]
α-terpineol	0.07 ± 0.01 [b]	Nd	0.04 ± 0.01 [a]
β-damascenone	0.58 ± 0.13 [a]	0.60 ± 0.07 [a]	1.65 ± 0.26 [b]
trans-geraniol	0.83 ± 0.03 [a]	1.10 ± 0.09 [b]	1.41 ± 0.43 [c]
geranic acid	4.45 ± 0.62 [a]	5.96 ± 0.21 [b]	3.82 ± 0.10 [a]
3-oxo-α-ionol	1.77 ± 0.09 [a]	4.00 ± 1.13 [b]	2.27 ± 0.51 [a]
3-hydroxy-7,8-dihydro-β-ionol	0.80 ± 0.09 [a]	2.00 ± 0.35 [b]	1.13 ± 0.02 [a]
Σ Terpenes and norisoprenoids	8.50 ± 0.81 [a]	13.65 ± 1.69 [b]	10.33 ± 1.19 [c]
benzaldehyde	0.70 ± 0.02 [b]	0.19 ± 0.03 [a]	0.63 ± 0.17 [b]
guaiacol	9.09 ± 1.47 [b]	0.80 ± 0.09 [a]	16.20 ± 5.01 [c]
benzyl alcohol	43.46 ± 2.44 [c]	15.29 ± 0.73 [b]	6.70 ± 0.57 [a]
4-vinylguaicol	15.48 ± 0.26 [a]	20.46 ± 0.24 [a]	28.90 ± 4.47 [b]
syringol	116.54 ± 18.60 [a]	112.95 ± 17.39 [a]	165.14 ± 34.44 [a]
benzoic acid	6.72 ± 0.16 [a]	7.09 ± 1.02 [a]	10.80 ± 0.89 [b]
vanillin	1.77 ± 0.07 [a]	2.12 ± 0.29 [b]	1.67 ± 0.02 [a]
methyl vanillate	6.82 ± 1.12 [a]	5.53 ± 1.05 [a]	10.51 ± 0.97 [b]
ethyl vanillate	nd	14.16 ± 1.41	nd
Σ Benzenic compounds	200.57 ± 19.66 [a]	178.58 ± 20.84 [a]	240.56 ± 41.44 [a]

Values with different superscripts in the same row denoted significant differences according to the Student–Newman–Keuls test at $p < 0.05$. Samples were defined according to the treatment applied: "Control wine": no treatment applied; "MW wine": microwave treatment at 700 W, 12 min of grape crushed; "MW wine without SO_2": microwave treatment at 700 W, 12 min of grape crushed elaborated without SO_2. nd: not detected.

C6 alcohols are one of the main components of the free fraction, with 1-hexanol as the major compound (Table 1). During the winemaking, C6 alcohols can suffer more modifications, especially in the maceration phase in red wines, in which the release of enzymes and precursors from the grape skin continues.

Wines from MW-treated musts presented higher quantities in total C6 alcohols, especially 1-hexanol, indicating their release from the grape skin cells during fermentation. However, a smaller increase was observed in those fermented wines without SO_2, which was possibly because oxidation reactions were favored in these conditions. The higher extraction of C6 alcohols can cause green and herbaceous aromas in wines, so it is not desirable to exceed its olfactory thresholds in wines [4].

The bound fraction of C6 alcohols was minority as in the must, and the effect of microwave maceration was scarce (Table 2). Some authors have also observed an increase in C6 alcohols in red wines from grapes undergoing sonication treatment during grape maceration [15,16].

Terpenes and norisoprenoids are compounds of sensorial relevance as they can contribute in an individual or synergistic way to the overall aroma of wines. The free fraction of these compounds was overall quantitatively important in all wines (Table 1), but an increase in wines from microwave maceration was observed, although the absence of SO_2 also decreased its concentration, except in the case of linalool. It has been described that many terpenes are very sensitive to oxidative processes, so the absence of SO_2 has been able to cause a loss of protection against oxidation [32].

Monoterpenes such as linalool, α-terpineol, and geraniol have floral or fruity aromas and low perception thresholds. Geranial was the majority terpene in wines from microwave maceration; this compound could have been formed by geraniol oxidation.

The most relevant compound within the norisoprenoids was the β-damascenone (floral, sweet, and honey-like aroma), which appeared in all samples at concentrations above its olfactory detection threshold (0.05 μg/L) [33]. Control wines had greater concentration, showing a possible degradation by MW treatment.

The glycosidically bound fraction of terpenes and norisoprenoids present in the wines was very small compared to the free fraction, which assumes that during the fermentation process, an important release of its precursors has occurred thanks to the glycosidic activity of yeasts [34,35]. MW treatment showed only a slight increase of this fraction in samples.

The effect of different maceration techniques on the extraction of terpenes and norisoprenoids has been studied by several authors, although the results depend on the conditions of the treatment applied. Tartian et al. [27] obtained higher amounts of monoterpenes in wines from cryomaceration than in those treated with ultrasound or microwave. Must thermovinification negatively affected the terpene content in wines [14]. In Monastrell wines from sonicated grapes, no changes in terpenes or norisoprenoids were observed [11,17].

Finally, in the group of benzenic compounds, we found quantitatively important compounds in both free and bound fractions (Tables 1 and 2). It should be noted that although they come from grapes, some of them can also be formed by the metabolism of yeasts, such as 4-vinylguaiacol or benzyl alcohol. MW-treated wines showed no significant differences in total benzenic compounds from control wines in both free and bound fraction.

Some sensory-relevant compounds such as guaiacol (smoky, sweet, medicinal aroma) and 4-vinylguaiacol (spicy aroma) were found in amounts above their perception thresholds in samples. Vanillin and its derivatives can jointly influence the vanilla aroma of wines. However, they all had lower concentrations in wines from microwave maceration and were unaffected by the absence of SO_2.

Benzenic compounds increased considerably in musts treated with microwaves, both in the free and bound fractions. However, in the wines, the differences due to MW treatment were smaller. During the red wine fermentation, grape skins' extraction of compounds was also possible slowly in the control wines, while MW treatment caused a much faster release of compounds at the beginning of fermentation. In the final wine, the quantities are balanced. Carew et al. [23] observed that the use of MW can favor the rapid release of phenolic compounds when shorter maceration processes are required in early press of Pinot Noir wines.

Similarly, the use of ultrasounds in red grape maceration was not effective in the extraction of benzenic compounds in red wines of different varieties [16,17], while grapes maceration with pulsed electric fields reduced the concentration of volatile phenols in white wines [8]. These authors justify this fact by a decrease of the hydroxycynamic acids precursors in the must via oxidation or because the enzymes involved may be affected.

3.3. Effect of Microwave Treatment on Volatile Compounds Formed during Alcoholic Fermentation with and without SO_2.

Volatile compounds formed during alcoholic fermentation form the basis of the wine aroma, especially in young wines from neutral grapes [36]. Grape maceration with MW has been shown to increase the amounts of phenolic compounds, amino acids, and other nutrients that can be used by yeasts as precursors of volatile compounds [24,25]. Likewise, the absence of SO_2 can influence the selection of yeasts that carry out fermentation by favoring the development of native yeasts and influencing the implantation of the starter, which could modify the profile of metabolites formed during fermentation.

Figure 2 shows the monitoring of wines fermentation, representing the average weight loss value of triplicates. As can be seen, the lag phase was longer in the case of control wine, since the weight loss of the flasks started later, while microwave-treated wines began fermentation more quickly. Wines from MW maceration also showed faster fermentation kinetics and higher fermentation yield, which was slightly higher in the absence of SO_2; this may be due to the lack of inhibition of yeasts due to SO_2. This effect has also been observed in wines from Pinot Noir elaborated with MW maceration at high potency [24,25] and with other physical treatments used in grape maceration such as sonication [11]. Treatment with MW, as well as sonication, could achieve a greater extraction of compounds from the inside of the grape such as amino acids, fatty acids, and other nutrients that the yeasts could use for their growth, producing a greater quantity of metabolites and accelerating the fermentation process.

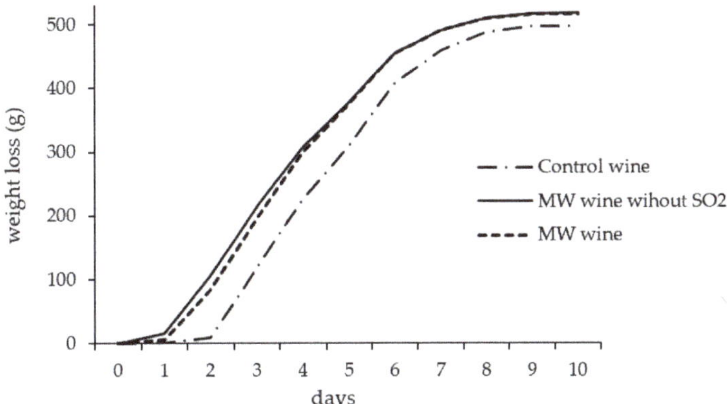

Figure 2. Alcoholic fermentation kinetics indicated by means weight loss in control wine and wines from microwave maceration elaborated with and without SO_2. "Control wine": no treatment applied; "MW wine": microwave treatment at 700 W, 12 min of grape crushed; "MW wine without SO_2": microwave treatment at 700 W, 12 min of grape crushed and elaborated without SO_2.

Table 3 shows the main volatile compounds produced during the alcoholic fermentation. Higher concentrations of most alcohols were identified in wines elaborated from microwave-treated grapes. This fact is justified by a greater extraction of amino acids and other volatile precursors and a more efficient fermentation due to MW treatment, which causes a greater production of metabolites by yeasts during alcoholic fermentation. The absence of SO_2 resulted in an increase in some major alcohols such as 2-methyl-1-propanol and 2-phenylethanol, which was characterized by its aroma of roses. It has been shown that the low availability of oxygen in fermentations with SO_2 can affect yeast metabolism by influencing the formation of alcohols and esters [37].

Table 3. Fermentative volatile compound concentration (µg/L) in control wines and those obtained from microwave-treated grapes and elaborated with and without SO_2 (n = 3).

Volatile Compounds	Control Wine	MW Wine	MW Wine without SO_2
	Mean ± SD	Mean ± SD	Mean ± SD
2-methyl-1-propanol	38.68 ± 4.74 [a]	154.91 ± 14.26 [b]	245.16 ± 9.03 [c]
butanol	13.16 ± 5.41 [a,b]	16.17 ± 0.32 [b]	7.47 ± 0.01 [a]
3-methyl-1-pentanol	4.13 ± 0.67 [a]	33.20 ± 6.28 [c]	15.09 ± 3.61 [b]
4-methyl-1-pentanol	17.58 ± 0.79 [a]	102.74 ± 21.09 [b]	26.80 ± 4.26 [a]
3-octanol	17.75 ± 3.57 [b]	1.85 ± 0.48 [a]	3.69 ± 0.40 [a]
1-octen-3-ol	2.55 ± 0.15 [b]	1.66 ± 0.31 [a]	4.01 ± 0.52 [c]
1-heptanol	2.12 ± 0.39 [a]	25.25 ± 0.45 [c]	11.90 ± 3.24 [b]
1-octanol	1.41 ± 0.18 [a]	4.30 ± 0.19 [b]	5.85 ± 1.33 [b]
Σ Alcohols	97.38 ± 3.12 [a]	340.07 ± 28.58 [b]	319.96 ± 2.22 [b]
isobutyric acid	53.62 ± 7.38 [a]	64.50 ± 3.78 [a]	57.53 ± 5.72 [a]
butanoic acid	5.87 ± 0.18 [a]	36.47 ± 1.30 [c]	13.10 ± 0.31 [b]
isovaleric acid	435.90 ± 48.07 [a]	474.04 ± 13.00 [a]	438.50 ± 25.45 [a]
pentanoic acid	2.84 ± 0.44 [a]	19.65 ± 1.81 [b]	3.26 ± 0.14 [a]
octanoic acid	1020.51 ± 19.23 [b]	1296.23 ± 48.52 [c]	621.99 ± 12.47 [a]
dodecanoic acid	57.93 ± 6.19 [a]	66.54 ± 11.11 [a]	78.01 ± 5.93 [a]
Σ Acids	1576.66 ± 74.02 [b]	1957.43 ± 56.69 [c]	1212.39 ± 20.09 [a]
ethyl butyrate	12.62 ± 1.09 [a]	61.87 ± 1.02 [c]	40.84 ± 0.77 [b]
ethyl isovalerate	4.14 ± 0.43 [a]	10.32 ± 2.44 [b]	15.00 ± 0.50 [c]
isoamyl acetate	1437.24 ± 53.61 [a]	2066.95 ± 41.66 [b]	946.51 ± 12.57 [a]
ethyl hexanoate	217.70 ± 38.39 [a]	327.35 ± 37.71 [b]	236.58 ± 15.79 [a]
ethyl piruvate	3.42 ± 0.20 [a]	7.76 ± 0.01 [b]	8.11 ± 1.12 [b]
hexyl acetate	1.88 ± 0.52 [a]	18.15 ± 3.29 [b]	5.25 ± 1.06 [a]
ethyl octanoate	262.41 ± 26.43 [b]	285.79 ± 15.27 [b]	84.76 ± 20.38 [a]
ethyl decanoate	22.68 ± 4.74 [a]	114.82 ± 9.64 [b]	33.76 ± 5.83 [a]
ethyl succinate	473.01 ± 5.05 [b]	443.08 ± 15.26 [b]	273.41 ± 73.23 [a]
ethyl 2-(OH)-4-methylpentanoate	41.38 ± 1.29 [a]	32.22 ± 0.98 [a]	36.59 ± 9.42 [a]
ethyl 3-(OH)-hexanoate	7.74 ± 0.31 [b]	7.45 ± 1.21 [b]	1.40 ± 0.38 [a]
ethyl 4-(OH)-butanoate	6594.31 ± 155.03 [c]	3116.845 ± 234.87 [b]	1991.30 ± 412.43 [a]
Σ Esters	9078.54 ± 217.17 [c]	6492.62 ± 222.50 [b]	3673.50 ± 482.15 [a]
2-phenylethyl acetate	76.36 ± 7.55 [a]	425.68 ± 15.83 [c]	175.51 ± 47.70 [b]
2-phenylethanol	24303.25 ± 555.69 [a]	35620.5 ± 2701.47 [b]	43608.55 ± 1702.30 [e]
4-vinyl-phenol	311.78 ± 24.28 [a]	417.51 ± 15.50 [b]	431.89 ± 39.91 [b]
tyrosol	3485.19 ± 839.37 [c]	2411.07 ± 7.49 [b]	1284.24 ± 374.96 [a]
benzeneacetic acid	10.19 ± 2.34 [a]	24.20 ± 0.81 [b]	39.38 ± 5.46 [c]
(2-phenylethyl) acetamide	56.51 ± 7.36 [a]	49.32 ± 5.69 [a]	99.23 ± 15.67 [b]
Σ Benzenic compounds	28243.26 ± 1370.95 [a]	38948.27 ± 2684.29 [b]	45638.81 ± 1402.11 [c]
3-(methylthio)-1-propanol	105.56 ± 13.62 [c]	38.72 ± 4.69 [a]	78.50 ± 10.71 [b]
3-(2H)-thiophenone, dihydro-2-methyl	204.19 ± 15.73 [c]	136.46 ± 2.75 [b]	50.87 ± 7.30 [a]
γ-butyrolactone	646.20 ± 50.68 [b]	49.33 ± 14.33 [a]	27.55 ± 4.40 [a]
γ-nonalactone	10.28 ± 1.35 [a]	15.29 ± 2.23 [a]	15.83 ± 3.99 [a]
pantolactone	13.26 ± 1.63 [b]	6.87 ± 2.36 [a]	14.09 ± 2.40 [b]
γ-decalactone	0.43 ± 0.03 [a]	2.05 ± 0.07 [c]	0.84 ± 0.13 [b]
Σ Furans & sulfur compounds	979.92 ± 60.10 [b]	248.72 ± 13.59 [a]	187.69 ± 9.18 [a]

Values with different superscripts in the same row denoted significant differences according to the Student–Newman–Keuls test at $p < 0.05$. Samples were defined according to the treatment applied: "Control wine": no treatment applied, 7 days of maceration; "MW wine": microwave treatment at 700 W, 12 min of grape crushed, 7 d of maceration; "MW wine without SO_2": microwave treatment at 700 W, 12 min of grape crushed, 7 d of maceration, elaborated without SO_2.

Fatty acids were most abundant in wines with MW maceration fermented with SO_2, with isovaleric acid and octanoic acid being the majority in all samples. Treatment with MW could favor the extraction of precursors of these acids from grape as unsaturated long chain fatty acids. Fatty acids have no pleasant aromas (rancid, cheese, or fatty), although they are usually found in wine below their olfactory thresholds. However, they are precursors of the fatty acid esters, among which we find compounds of great sensory relevance in young wines [36].

The amount of fatty acid esters in microwave-macerated wines was significantly higher than in control wines, although the absence of SO_2 caused the decrease of some of them (ethyl octanoate, ethyl decanoate), which is consistent with its content of fatty acids since they share the metabolic pathway from Acetyl-CoA. Short-chain fatty acid ethyl esters with low olfactory thresholds offer wines fruity notes. Ethyl butyrate (with strawberries, apple aroma) and ethyl hexanoate (with fruity, green apple aroma) were found in all wines above their odor thresholds [33]. Ethyl isovalerate characterized by its apple, "sweet aroma" with a low odor threshold (3 μg/L) [38] was more abundant in MW-treated wines elaborated without SO_2.

As for acetates, isoamyl acetate (with "banana aroma") and 2-phenyl ethyl acetate ("roses", "floral" aroma) were the most quantitatively and sensory important compounds. In both cases, the MW macerated wines in the presence of SO_2 obtained the highest concentrations, although the olfactory thresholds were exceeded in all samples [33]. Garde-Cerdan and Ancin-Azpilicueta [37] did not observe differences in the amounts of esters and acetates in vinifications without SO_2, except for the ethyl hexanoate that increased. In our case, a significant decrease was observed, which indicates the influence of oxygen availability and must composition on the production of metabolites by yeasts.

Within the hydroxyacids, the most abundant was the 4-OH-ethyl butyrate that presented the highest concentration in the control wine. However, the ethyl 2-hydroxy-4-methylpentanoate related with the blackberry aroma in red wines was the most significant due to its lower odor thresholds [39] and did not undergo modifications due to MW treatment.

Benzenic compounds include 4-vinylphenol synthesized by yeasts from the cinnamic acids of the grape, which justifies that a greater extraction of these acids increases their concentration in MW-treated wines. Due to its unpleasant odor (medicinal, pharmaceutical), it is not advisable that its concentration be increased in wines. 2-Phenylethanol and tyrosol belong to the group of major alcohols that are formed from certain precursor amino acids. A greater extraction of amino acids due to microwave treatment, as well as the greater fermentative activity observed in these wines would justify their greater presence in the wines treated with MW.

Sulfur compounds (methionol and 3(2H) dihydro 2-methyl thiophenone) were more abundant in control wines. Methionol is very sensitive to oxidation, forming methional with a lower detection threshold; however, its decrease in the absence of SO_2 has been observed, which is probably due to less sulfur input. [40].

Lactones usually contribute pleasant aromas to the wine. Of the lactones identified, the γ-butyrolactone was the majority, being more abundant in the control wines. However, other lactones such as γ-nonalactone (coconut-like aroma) and γ-decalactone showed higher concentrations in wines with MW maceration.

In general, wines from microwave-treated grapes had higher quantities of most fermentation compounds, especially some esters, acetates, and alcohols of sensory interest. However, the absence of SO_2 resulted in a decrease in the concentration of some of these compounds, which could be justified by changes generated by oxidative processes or due to different metabolic routes used by yeasts depending on the availability of oxygen.

The works about the use of other alternative techniques in grape maceration such as sonication showed a variable behavior on its effect in volatile fermentation compounds, depending on the grape variety and the conditions used (maceration time, power used, etc.). Lower production of acetates, increasing of total esters and alcohols, and no changes have been observed by different authors [15,16,41]. Geffroy et al. [14] observed that the heating temperature in thermovinification did not induce any changes in ethyl esters, acetates, and acids, despite the increase in yeast assimilable nitrogen (YAN) due to treatment.

3.4. Sensory Analysis of Control and Microwave-Treated Wine Elaborated with and without SO_2

Changes observed in volatile composition due to microwave treatment in grape maceration and absence of SO_2 may have sensory effects on wines. Therefore, an olfactory descriptive sensory analysis of wines was carried out. The olfactory attributes that best represented the samples selected by the panel of judges were red berry, prune, green, floral, and odor intensity.

Figure 3 shows in the form of a spider web the scores given by the tasters to these attributes in control wines and those treated with microwaves (with and without SO_2). All attributes showed significant differences between samples according to the ANOVA test. The wines from maceration with microwave were the ones that had the highest scores in the attributes floral and red berry odor, as well as a greater intensity in the aroma. This may be related to its higher content in esters and acetates fundamentally. Table 4 shows the odor active values (OAVs) of those compounds whose concentrations in the tested wines were closer to their perception thresholds, and therefore they may have a greater sensory impact. As can be seen, the microwave-treated wine presented the highest OAVs values in all fatty acid esters with descriptors related to fruit attributes that exceeded their detection threshold. However, some berry notes have been linked in red wines to certain profiles of ethyl esters, including odorants at subthreshold concentrations [42].

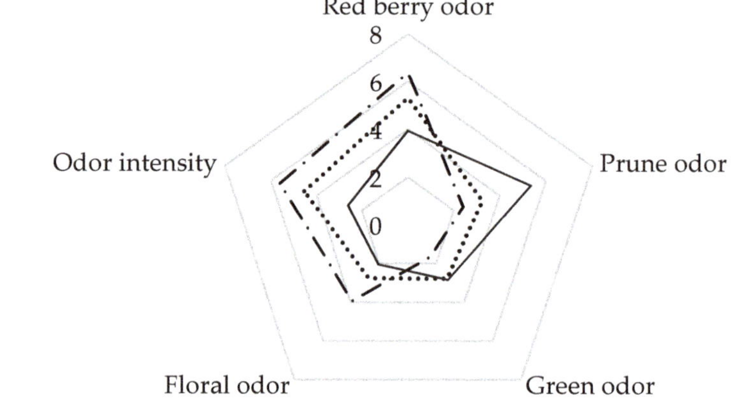

Figure 3. Olfactory attributes scores of control wines and wines from microwave maceration elaborated with and without SO_2. "Control wine": no treatment applied; "MW wine": microwave treatment at 700 W, 12 min of grape crushed; "MW wine without SO_2": microwave treatment at 700 W, 12 min of grape crushed, elaborated without SO_2.

Table 4. Odor-active values of the most sensorially significant volatile compounds in control wines and microwave-treated wines fermented with and without SO_2.

Volatile Compound	Odor Threshold µg/L *	Odor Descriptor	OAV **		
			Control Wine	MW Wine	MW Wine without SO_2
linalool	15	Citrus, floral, sweet	0.4	0.4	0.8
β-damascenone	0.05	Honey, sweet,	1459.4	515.8	462.4
guaiacol	10	Smoke, sweet, medicine	4.6	2.6	5.5
vinylguaicol	40	Spices, curry	1.8	0.7	1.6
ethyl butyrate	20	Fruity, strawberry, sweet,	0.6	3.1	2.0
ethyl isovalerate	3	Apple, sweet	1.4	3.4	5.0
isoamyl acetate	30	Banana, fruity, sweet	47.9	68.9	31.5
ethyl hexanoate	5	Fruity, green, apple.	43.5	65.5	47.3
ethyl octanoate	5	Sweet, fruity, pear	52.5	57.2	16.9
2-phenylethyl acetate	250	Flowery	0.3	1.7	0.7
2-phenylethanol	10,000	Rose, honey	2.4	3.5	4.3
γ-nonalactone	25	soft coconut, sweet	0.4	0.6	0.6

* odor thresholds values have been obtained from the References: [33,38]. ** OAVs were calculated dividing the mean concentration of each compound in wines by its odor threshold.

Scores for floral and red berry attributes were also higher for wines made without SO_2, which may be related to the higher OAVs in compounds such as linalool (with a OAV close to 1), ethyl isovalerate with apple, sweet aroma, and 2-phenylethanol, characterized by its rose aroma, in these wines. Other authors have described an increase in the notes related to floral and fruity aroma in wines elaborated with different SO_2 substitutes, suggesting that the presence of SO_2 could mask and neutralize these aromas [43,44].

Prune aroma was identified by tasters in greater quantities in control wines. This aroma has been linked to the presence of high levels of β-damascenone and γ-nonalactone [45]. The first was the compound with the highest odor active value in all samples, but the control wine presented the highest values.

The green odor attribute may be related to the presence of C6 alcohols, which were overall superior in microwave-treated wines due to their greater extraction of the grape skin. Therefore, it is logical to think that this attribute will increase in these wines, although their scores were not excessively high; in fact, none of them exceeded their olfactory threshold individually. This fact is common in wines undergoing intense maceration treatments, such as sonication or cryomaceration [16].

The effect of other grape maceration techniques on the sensory perception of wines differs according to the studies carried out. Maceration with ultrasound and thermovinification increased floral and fruity scores, obtaining greater aromatic complexity in Syrah and Monastrell wines [15,16,46]. No defect or unpleasant aromas were detected in the wines analyzed, which has been described with other intensive maceration techniques [47]. Some compounds such as guaiacol or vinylguaicol that presented OAVs higher than 1, which is related to a spicy or medicinal aroma, were not detected.

4. Conclusions

The microwave treatment with medium intensity (700 W) and temperature control applied in the maceration of grapes crushed at laboratory scale increased the amounts of varietal compounds of the must in a very evident way in both the free and the glycosidically bound fractions. This increase may be due to a greater extraction of these compounds from the grape skin thanks to MW treatment.

Wines from MW maceration in the presence of SO_2 obtained higher amounts of C6 alcohols, terpenes, and norisoprenoids in free form, while there were few changes in the bound fraction and in the benzenic compounds concentration. On the other hand, wines with MW treatment showed faster fermentation kinetics and shorter lag phase, resulting in an increase in some volatile compounds from fermentation of sensory relevance such as alcohols, esters, and acetates. The absence of sulfurous in treated MW wines resulted in changes in the concentrations of some volatile compounds, such as a decrease in some esters or an increase in linalool or 2-phenylethanol.

In addition, the wines best valued by the tasters for their greater red berry and floral odor and intensity aroma were those treated with MW, and especially those elaborated without SO_2, which shows that treatment with MW can be very suitable to increase the aromatic potential of wines by reducing SO_2 levels in their production.

Supplementary Materials: The following are available online: https://www.mdpi.com/article/10.3390/foods10061164/s1 Table S1. Free volatile compound concentrations (µg/L) in control must and that obtained from microwave-treated grapes (MW). Table S2. Glycosidically bound volatile compound concentrations (µg/L) in control must (C) and that obtained from microwave-treated grapes (MW).

Author Contributions: Conceptualization, M.S.P.-C.; Formal analysis, R.M.G. and R.O.S.; Funding acquisition, M.C.D.-M. and M.S.P.-C.; Project administration, M.C.D.-M. and M.S.P.-C.; Supervision, M.C.D.-M. and M.S.P.-C.; Writing—original draft, R.M.G. and R.O.S.; Writing—review and editing, M.E.A.P. and M.S.P.-C. All authors have read and agreed to the published version of the manuscript.

Funding: This research was funded by the Ministerio de Ciencia, Innovación y Universidades (Spanish Government and Feder Funds), grant number RTI2018-093869-B-C22.

Data Availability Statement: Not available.

Conflicts of Interest: The authors declare no conflict of interest.

References

1. Clodoveo, M.L.; Dipalmo, T.; Rizzello, C.G.; Corbo, F.; Crupi, P. Emerging technology to develop novel red winemaking practices: An overview. *Innov. Food Sci. Emerg. Technol.* **2016**, *38*, 41–56. [CrossRef]
2. Cui, Y.; Lv, W.; Liu, J.F.; Wang, B.J. Effect of different ending fermentation technologies on microbial-stability of Italian Riesling low alcohol sweet white wine. *Adv. Mater. Res.* **2012**, *393*, 1165–1168. [CrossRef]
3. Rougier, C.; Prorot, A.; Chazal, P.; Leveque, P.; Lepratb, P. Thermal and Nonthermal Effects of Discontinuous Microwave Exposure (2.45 Gigahertz) on the Cell Membrane of Escherichia coli. *Appl. Environ. Microbiol.* **2014**, *80*, 4832–4841. [CrossRef] [PubMed]
4. Baumes, R. Wine aroma precursors. In *Wine Chemistry and Biochemistry*; Moreno Arribas, M.V., Polo, M.C., Eds.; Springer Science & Business Media LLD: New York, NY, USA, 2009; pp. 251–274. [CrossRef]
5. Hjelmeland, A.K.; Ebeler, S.E. Glycosidically Bound Volatile Aroma Compounds in Grapes and Wine: A Review. *Am. J. Enol. Vitic.* **2015**, *66*, 1–11. [CrossRef]
6. Sacchi, K.L.; Bisson, L.F.; Adams, D.O. A review of the effect of winemaking techniques on phenolic extraction in red wines. *Am. J. Enol. Vitic.* **2005**, *56*, 197–206. [CrossRef]
7. Atanackovic, M.; Petrovic, A.; Jovic, S.; Gojkovic-Bukarica, L.; Bursac, M.; Cvejic, J. Influence of winemaking techniques on the resveratrol content, total phenolic content and antioxidant potential of red wines. *Food Chem.* **2012**, *131*, 513–518. [CrossRef]
8. Comuzzo, P.; Marconi, M.; Zanella, G.; Querzè, M. Pulsed electric field processing of white grapes (cv. Garganega): Effects on wine composition and volatile compounds. *Food Chem.* **2018**, *264*, 16–23. [CrossRef]
9. Jiranek, V.; Grbin, P.; Yap, A.; Barnes, M.; Bates, D. High power ultrasonics as a novel tool offering new opportunities for managing wine microbiology. *Biotechnol. Lett.* **2008**, *30*, 1–6. [CrossRef]
10. García Martín, J.F.; Sun, D.W. Ultrasound and electric fields as novel techniques for assisting the wine ageing process: The state-of-the-art research. *Trends Food Sci. Technol.* **2013**, *33*, 40–53. [CrossRef]
11. Pérez-Porras, P.; Bautista-Ortín, B.; Jurado, R.; Gómez-Plaza, E. Using high-power ultrasounds in red winemaking: Effect of operating conditions on wine physico-chemical and chromatic characteristics. *LWT-Food Sci. Technol.* **2021**, *138*, 110645–110653. [CrossRef]
12. Parenti, A.; Spugnoli, P.; Calamai, L.; Ferrari, S.; Gori, C. Effects of cold maceration on red wine quality from Tuscan Sangiovese grape. *Eur. Food Res. Technol.* **2004**, *218*, 360–366. [CrossRef]
13. Geffroy, O.; Lopez, R.; Serrano, E.; Dufourcq, T.; Gracia-Moreno, E.; Cacho, J.; Ferreira, V. Changes in analytical and volatile compositions of red wines induced by pre fermentation heat treatment of grapes. *Food Chem.* **2015**, *187*, 243–253. [CrossRef] [PubMed]
14. Geffroy, O.; Lopez, R.; Feilhes, C.; Violleau, F.; Kleiber, D.; Favarel, J.L.; Ferreira, V. Modulating analytical characteristics of thermovinified Carignan musts and the volatile composition of the resulting wines through the heating temperature. *Food Chem.* **2018**, *257*, 7–14. [CrossRef]
15. Ruiz-Rodríguez, A.; Carrera, C.; Palma-Lovillo, M.; García Barroso, C. Ultrasonic treatments during the alcoholic fermentation of red wines: Effects on 'Syrah' wines. *Vitis* **2019**, *58*, 83–88. [CrossRef]

16. Oliver Simancas, R.; Díaz-Maroto, M.C.; Alañón Pardo, M.E.; Pérez Porras, P.; Bautista Ortín, A.B.; Gómez-Plaza, E.; Pérez-Coello, M.S. Effect of power ultrasounds treatment on free and glycosidically-bound volatile compounds and sensorial profile of red wines. *Molecules* **2021**, *26*, 1193. [CrossRef] [PubMed]
17. Bautista-Ortín, A.B.; Jiménez-Martínez, M.D.; Jurado, R.; Iniesta, J.A.; Terrades, S.; Andrés, A.; Gómez-Plaza, E. Application of high-powder ultrasounds during red wine vinification. *Int. J. Food Sci. Technol.* **2017**, *52*, 1314–1323. [CrossRef]
18. Letellier, M.; Budzinski, H. Microwave assisted extraction of organic compounds. *Analysis* **1999**, *27*, 259–271. [CrossRef]
19. Kaufmann, B.; Christen, P. Recent extraction techniques for natural products: Microwave-assisted extraction and pressurised solvent extraction. *Phytochem. Anal.* **2002**, *13*, 105–113. [CrossRef]
20. Liazid, A.; Guerrero, R.F.; Cantos, E.; Palma, M.; Barroso, C.G. Microwave assisted extraction of anthocyanins from grape skins. *Food Chem.* **2011**, *124*, 1238–1243. [CrossRef]
21. Bittar, S.A.; Perino-Issartier, S.; Dangles, O.; Chemat, F. An innovative grape juice enriched in polyphenols by microwave-assisted extraction. *Food Chem.* **2013**, *141*, 3268–3272. [CrossRef]
22. Thostenson, E.T.; Chou, T.W. Microwave processing: Fundamentals and applications. *Compos. Part A* **1999**, *30*, 1055–1071. [CrossRef]
23. Carew, A.L.; Close, D.C.; Dambergs, R.G. Yeast strain affects phenolic concentration in Pinot noir wines made by microwave maceration with early pressing. *J. Appl. Microbiol.* **2015**, *118*, 1385–1394. [CrossRef] [PubMed]
24. Carew, A.L.; Gill, W.; Close, D.C.; Dambergs, R.G. Microwave maceration with early pressing improves phenolics and fermentation kinetics in Pinot noir. *Am. J. Enol. Vitic.* **2014**, *65*, 401–406. [CrossRef]
25. Carew, A.L.; Sparrow, A.M.; Curtin, C.D.; Close, D.C.; Dambergs, R.G. Microwave maceration of Pinot Noir grapemust: Sanitation and extraction effects and wine phenolics outcomes. *Food Bioproc. Technol.* **2014**, *7*, 954–963. [CrossRef]
26. Casassa, L.F.; Sari, S.E.S.; Bolcato, E.A.; Fanzone, M.L. Microwave-Assisted Extraction Applied to Merlot Grapes with Contrasting Maturity Levels: Effects on Phenolic Chemistry and Wine Color. *Fermentation* **2019**, *5*, 15. [CrossRef]
27. Tartian, A.C.; Cotea, V.V.; Niculaua, M.; Zamfir, C.I.; Colibaba, C.L.; Moros, A.M. The influence of the different techniques of maceration on the aromatic and phenolic profile of the Busuioaca de Bohotin wine. BIO Web of Conferences. In Proceedings of the 40th World Congress of Vine and Wine, Sofia, Bulgaria, 29 May–2 June 2017.
28. Jankovic, S.M.; Milosev, M.Z.; Novakovic, M.L.J. The effects of microwave radiation on microbial culture. *Hosp. Pharmacol. Int. Multidiscip. J.* **2014**, *1*, 102–108. [CrossRef]
29. González-Arenzana, L.; Santamaría, P.; López, R.; Garijo, P.; Gutiérrez, A.R.; Garde-Cerdán, T.; López-Alfaro, I. Microwave technology as a new tool to improve microbial control of oak barrels: A preliminary study. *Food Control* **2013**, *30*, 536–539. [CrossRef]
30. Sánchez Palomo, E.; González-Viñas, M.A.; Díaz-Maroto, M.C.; Soriano Pérez, A.; Pérez-Coello, M.S. Aroma potential of Albillo wines and effect of skin-contact treatment. *Food Chem.* **2007**, *103*, 631–640. [CrossRef]
31. Roman, T.; Tonidandel, L.; Nicolini, G.; Bellantuono, E.; Barp, L.; Larcher, R.; Celotti, E. Evidence of the possible interaction between ultrasound and thiol precursors. *Foods* **2020**, *9*, 104. [CrossRef]
32. Bayonove, C.; Baumes, R.; Crouzet, J.; Günata, Z. Aromas. In *Enología: Fundamentos Científicos y Tecnológicos*; Flanzy, C., Ed.; AMV Ediciones, Mundi Prensa: Madrid, Spain, 2000; pp. 137–176.
33. Guth, H. Quantification and sensory studies of character impact odorants of different white varieties. *J. Agric. Food Chem.* **1997**, *45*, 3027–3032. [CrossRef]
34. Ugliano, M.; Bartowsky, E.J.; Mccarthy, J.; Moio, L.; Henschke, P.A. Hydrolysis and transformation of grape glycosidically bound volatile compounds during fermentation with three Saccharomyces yeast strains. *J. Agric. Food Chem.* **2006**, *54*, 6322–6331. [CrossRef] [PubMed]
35. Loscos, N.; Hernández-Orte, P.; Cacho, J.; Ferreira, V. Release and formation of varietal aroma compounds during alcoholic fermentation from nonfloral grape odorless flavor precursors fractions. *J. Agric. Food Chem.* **2007**, *55*, 6674–6684. [CrossRef] [PubMed]
36. Ferreira, V. *Managing Wine Quality*; Volume 2: Oenology and Wine Quality; Reynolds, A.G., Ed.; Woodhead Publishing Limited: Cambridge, UK, 2010; pp. 3–28. Volatile aroma compounds and wine sensory attributes. [CrossRef]
37. Garde-Cerdan, T.; Ancin-Azpilicueta, C. Effect of SO_2 on the formation and evolution of volatile compounds in wines. *Food Control* **2007**, *18*, 1501–1506. [CrossRef]
38. Ferreira, V.; Lopez, R.; Cacho, J.F. Quantitative determination of the odorants of young red wines from different grape varieties. *J. Sci. Food Agric.* **2000**, *80*, 1659–1667. [CrossRef]
39. Campo, E.; Cacho, J.; Ferreira, V. Multidimensional chromatographic approach applied to the identification of novel aroma compounds in wine. Identification of ethyl cyclohexanoate, ethyl 2-hydroxy-3-methylbutyrate and ethyl 2-hydroxy-4-methylpentanoate. *J. Chromatogr. A* **2006**, *1137*, 223–230. [CrossRef]
40. Raposo, R.; Ruiz-Moreno, M.J.; Garde-Cerdan, T.; Puertas, B.; Moreno-Rojas, J.M.; Gonzalo-Diago, A.; Cantos-Villar, E. Effect of hydroxytyrosol on quality of sulfur dioxide-free red wine. *Food Chem.* **2016**, *192*, 25–33. [CrossRef]
41. Martínez-Pérez, M.P.; Bautista-Ortín, A.B.; Pérez-Porras, P.; Jurado, R.; Gómez-Plaza, E. A new approach to the reduction of alcohol content in red wines: The use of high-power ultrasounds. *Foods* **2020**, *9*, 726. [CrossRef]

42. De la Fuente-Blanco, A.; Sáenz-Navajas, M.P.; Valentin, V.; Ferreira, V. Fourteen ethyl esters of wine can be replaced by simpler ester vectors without compromising quality but at the expense of increasing aroma concentration. *Food Chem.* **2020**, *307*, 125553–125564. [CrossRef]
43. Raposo, R.; Ruiz-Moreno, M.J.; Garde-Cerdan, T.; Puertas, B.; Moreno-Rojas, J.M.; Zafrilla, P.; Cantos-Villar, E. Replacement of sulfur dioxide by hydroxytyrosol in white wine: Influence on both quality parameters and sensory. *LWT Food Sci. Technol.* **2016**, *65*, 214–221. [CrossRef]
44. Sonni, F.; Cejudo Bastante, M.J.; Chinnici, F.; Natali, N.; Riponi, C. Replacement of sulfur dioxide by lysozyme and oenological tannins during fermentation: Influence on volatile composition of white wines. *J. Sci. Food Agric.* **2008**, *89*, 688–696. [CrossRef]
45. Pons, A.; Lavigne, V.; Eric, F.; Darriet, F.; Dubourdieu, D. Identification of volatile compounds responsible for prune aroma in prematurely aged red wines. *J. Agric. Food Chem.* **2008**, *56*, 5285–5290. [CrossRef] [PubMed]
46. Pozzatti, M.; Guerra, C.C.; Martins, G.; dos Santos, I.D.; Wagner, R.; Ferrão, M.F.; Manfroi, V. Effects of winemaking on 'Marselan' red wines: Volatile compounds and sensory aspects. *Ciência Téc. Vitiv.* **2020**, *35*, 63–75. [CrossRef]
47. Gracin, L.; Jambrak, A.R.; Juretic, H.; Dobrovic, S.; Barukcic, I.; Grozdanovic, M.; Smoljanic, G. Influence of high power ultrasound on Brettanomyces and lactic acid bacteria in wine in continuous flow treatment. *Appl. Acoust.* **2016**, *103*, 143–147. [CrossRef]

 *foods*

Article

A Comparative Study on Volatile Compounds and Sensory Profile of White and Red Wines Elaborated Using Bee Pollen versus Commercial Activators

Antonio Amores-Arrocha *, Pau Sancho-Galán , Ana Jiménez-Cantizano and Víctor Palacios

Department of Chemical Engineering and Food Technology, Vegetal Production Area, Faculty of Sciences, Agrifood Campus of International Excellence (ceiA3), University of Cadiz, 11510 Cadiz, Spain; pau.sancho@uca.es (P.S.-G.); ana.jimenezcantizano@uca.es (A.J.-C.); victor.palacios@uca.es (V.P.)
* Correspondence: antonio.amores@uca.es

Abstract: Lack of nutrients in grape may cause problems for a proper alcoholic fermentation process, resulting in an altered aromatic profile of the wines. To avoid this situation, commercial winemakers often use fermentation activators, which are usually combinations of ammonium salts, inactivated yeast and thiamine. In addition, it has been shown that bee pollen addition to the grape can help to improve fermentation, resulting in better volatile compound profile of wines responsible for sensory quality. For this reason, the aim of this research work was to carry out a comparative study using bee pollen versus commercial fermentation activators in white and red winemaking. The same dose of bee pollen and commercial activators (0.25 g/L) were used in all experiments. Volatile compounds were analyzed by gas chromatography–mass spectrometry, odor activity values were determined to assess odorant impact of various volatile compound families, and finally a descriptive sensory analysis was carried out. Then, the triangular test and the ranking assay were used to identify perceptible differences as well as preference among the wines elaborated. Compared to commercial activators, bee pollen wines increased volatile compound formation, mainly higher alcohols, esters, and terpenes, enhancing fruity and floral odorant series. On the other hand, triangular test showed significant differences between wines, and the ranking assay showed a greater preference for bee pollen wines.

Keywords: bee pollen; volatile compounds; fermentative activator; odorant activity value; alcoholic fermentation; white wines; red wines

Citation: Amores-Arrocha, A.; Sancho-Galán, P.; Jiménez-Cantizano, A.; Palacios, V. A Comparative Study on Volatile Compounds and Sensory Profile of White and Red Wines Elaborated Using Bee Pollen versus Commercial Activators. *Foods* **2021**, *10*, 1082. https://doi.org/10.3390/foods10051082

Academic Editors: Fulvio Mattivi and Matteo Bordiga

Received: 12 April 2021
Accepted: 11 May 2021
Published: 13 May 2021

Publisher's Note: MDPI stays neutral with regard to jurisdictional claims in published maps and institutional affiliations.

1. Introduction

Wine flavor is a combined perception of visual attributes, taste, and aroma, with the aroma being the most responsible for the global perception of wines [1]. The role of wine aroma can be a determining factor in consumer's preferences [2–4] and is one of the key aspects to be taken into account during winemaking. Wine aroma compounds can be grouped according to their origin: varietal aromas are found in grapes, fermentative aromas are derived from the alcoholic and malolactic fermentation, and ageing aromas are obtained during ageing or storage [5]. However, many of the volatile compounds generated during alcoholic fermentation (especially esters, higher alcohols, volatile acids, and various terpenoids and thiols), produced via the metabolic activity of *Saccharomyces cerevisiae*, account quantitatively for biggest fraction of the total aroma composition of wine [1,5,6].

Fermentative yeasts require adequate nutrient levels in the must for proper alcoholic fermentation. A lack of nutrients in grape musts could induce a rapid growth of non-target microorganisms and a displacement of the starter yeast strains could result in sluggish or stuck fermentations [7,8]. Nitrogen or certain vitamin (thiamine and pantothenic acid) deficiencies in grape must could induce problems during the alcoholic fermentation process, resulting in significant sensory defects in final wines [9–13].

In the current wine industry, the use of dehydrated yeast cultures (commercial active dry yeast, ADY) is an extended practice for winemaking [14]. Commercial yeast strains are usually implicitly linked to a high nutrient demand for proper inoculum implantation and subsequent alcoholic fermentation [15]. For a correct development of alcoholic fermentation, these yeasts need to be supplemented with mix of macronutrients, such as sugars, free amino nitrogen, phosphorus, potassium, and magnesium, and micronutrients, such as calcium, copper, iron, manganese, and zinc [14]. In this regard, commercial yeast activators are used in wineries to correct nutritional deficiencies in grape musts [16–20] in order to supply yeast nutritional necessities and to avoid the appearance of problems during alcoholic fermentation [21–23].

Bee pollen is a natural source of proteins, essential amino acids, lipids, fatty acids, sterols, phospholipids, carbohydrates, carotenoids, and polyphenols [24–30]. Previous published research works proposes the use of bee pollen as an alternative to the use of commercial fermentation activators. The effects of its use at different doses have been studied on the sensory profile of white [31] and red [32] wines, showing that bee pollen improves the sensory profile of wines when used at low doses. For this reason, this research study proposes a comparative study between bee pollen use versus commercial activators employed for white and red winemaking.

2. Materials and Methods

2.1. Experimental Procedure

The grape must of the white variety Palomino Fino was obtained from the Cooperativa Andaluza winery, Unión de Viticultores Chiclaneros of Chiclana de la Frontera (36.426067, −6.148189, 50 m above sea level) and the Riesling grape must from a private winery of Jerez de la Frontera (36.666594, −6.114847, 50 m above sea level). The red grape variety, Tintilla de Rota, was harvested from the private winery Luis Pérez in Jerez de la Frontera (36.700167, −6.192778, 100 m above sea level). All the vineyards were located in southern Andalusia, Cádiz (Spain), and were grown under warm climate conditions over an albariza (limestone) soil.

Once the white grape musts were obtained, they were added with 90 mg/L of potassium metabisulphite and racked for 24 h at controlled temperature (10 °C). Red grapes were destemmed and crushed, and the resulting paste was dosed with 25 mg/L $K_2O_5S_2$ (Sigma-Aldrich Chemical S.A., Madrid, Spain). The white grape must and the paste (pulp + skins) of Tintilla de Rota were placed in 5-liter glass fermenters jacketed for temperature control. For each vinification (bee pollen or commercial activator addition), 9 simultaneous fermenters were prepared including control. Each experiment was conducted in triplicate (n = 3).

Bee pollen has been widely employed by the research group in several studies, showing good results in both white [31,33,34] and red [32,35] vinification. For this reason, the dose of this comparative study was estimated as 0.25 g/L of bee pollen, previously crushed and preserved under dark and desiccation conditions. For white vinifications, a commercial activator SUPERSTART® BLANC (Laffort, Bordeaux, France) was used, with a specific formulation for white wine vinification conditions. For the red wine, SUPERSTART® ROUGE (Laffort, Bordeaux, France) was used, a preparation based on inactivated yeasts and autolysates of selected yeasts that is rich in vitamins, minerals, fatty acids, and sterols (er-gosterol) and adapted in particular for red wine production. In both cases (white and red), vinifications with commercial activators were carried out at a dose of 0.25 g/L, equal to the dose of bee pollen used.

Alcoholic fermentation (AF) was carried out using a commercial yeast starter *Saccharomyces cerevisiae* Lalvin 71B® (Lallemand, Barcelona, Spain) under controlled temperature conditions (20 °C). In the red wine vinification, once AF was finished, malolactic fermentation (MLF) was carried out using a commercial strain of a lactic acid bacteria (LAB) *Oenococcus oeni* S11B P2 Instant (Laffort, Bordeaux, France) at the recommended dosage (1 g/hL).

2.2. Aroma Compounds and Odorant Activity Values (OAV)

Major volatile compounds were determined by gas chromatography with flame ionization detection (GC-FID) using 4-methyl-2-pentanol as internal standard and standard calibration to determinate retention times and calibration curves. On the other hand, the minority volatiles were identified and quantified by semiquantitative analysis, using 1-heptanol (Sigma-Aldrich Química, S.A., Madrid, Spain) as internal standard, assuming a response factor equal to one according to the methodology described by Amores-Arrocha et al. [31]. To analyze the odorant activity value (OAV) on final wines, we calculated the ratio of the concentration of each compound and its perception threshold. The same methodology described by Amores-Arrocha et al. [31,32] and aroma descriptors previously published by several authors were employed for this purpose [16,18,20].

2.3. Sensory Testing

All tasting sessions were conducted in a temperature-controlled tasting room at the University Institute of Viticulture and AgriFood Research (IVAGRO, Puerto Real, Cádiz). Each taster was located in an individual booth with controlled lighting that was separated from the rest of the judges by panels to avoid possible interactions. Each judge was provided with the same amount of wine in standard ISO 3591 (1997) [36] glass cups covered with a glass lid, and temperature in the room was controlled in order to avoid the evaporation of volatile compounds (20 $\pm$ 2 °C). Each tasting session (descriptive analysis, sorting test, and triangular test) took place on different days.

2.3.1. Sensory Descriptive Analysis

A total of 20 judges, previously trained and experienced in sensory evaluations of white and red wines, were involved in the descriptive sensory analysis. All wine samples were randomly coded with three digits and presented disorderly. Each judge was assigned a tasting sheet to evaluate the intensity of different attributes on a scoring scale of 0–10 points. Attributes evaluated during the tasting sessions were previously selected, taking into account the tasting descriptors for white and red wines according to Jackson (2009) [37].

2.3.2. Classification Test

The ranking test is a sensory evaluation technique whereby a series of samples can be ordered according to established criteria. For this test, judges were instructed to order the wines samples by hedonic preference in order to determine the existence of differences between them. Once samples were classified, an increasing score was assigned according to the order of preference and the sums of the rankings of the samples were calculated for each judge, rejecting the null hypothesis (H_0) when there was no difference between the samples. Page's test value was applied for a comparison with respect to a known order. Observing the results pre-established by ISO 8587:2006 AENOR (2010) [38], we find that H_0 will be rejected if F test > F, considering the number of judges, the number of samples, and the risk assumed; therefore, it may be concluded if there are consistent differences between the ranking of the samples. A total of 29 judges were used to carry out the ranking test.

2.3.3. Triangular Test

To determine any perceptible variances between wines elaborated (control, pollen and commercial), we applied the ISO 4120:2007 AENOR (2008) [39] triangular test. This test allows us to know if there is any perceptible difference between three samples, regardless of the possible nature of these differences [40]. In the triangular test, triad samples are presented simultaneously, where two of them are always identical and one is different. Panelists must indicate in each case which sample is the different one. In this sense, it is hypothesized that the probability of choosing the different sample when there is no difference between them is 1/3 (H0: Pt = 1/3). In this modality, samples were presented to the judges in different position sequences (AAB, AAC, BCC, ABB, ACC, and BCC). After a judges' evaluation, the next round of samples was placed in turn in the order of the sequence, and therefore during the evaluation sessions, all samples were presented in equal number and randomly. According to the standard, if the number of correct answers is greater than or equal to the number established in the table (corresponding to the number of judges and the risk level in the test), it can be concluded that there are perceptible differences between samples. The triangular test was carried out with the participation of 28, 27, and 26 judges for Palomino Fino, Riesling, and Tintilla de Rota wines, respectively.

2.4. Data Analysis

Means and standard deviations were determined by two-way ANOVA analysis and Bonferroni's multiple range test (BSD), considering $p < 0.05$ as significant (GraphPad Prism version 6.01 for Windows, GraphPad Software, San Diego, CA, USA). All tests were performed in triplicate ($n = 3$). For principal component analysis (PCA), SPSS 24.0 statistical software package (SPSS Inc., Chicago, IL, USA) was employed.

3. Results and Discussion

3.1. Influence of Bee Pollen and Commercial Activators on Volatile Compounds

Tables 1–3 shows in comparative terms the effect of the addition of bee pollen, commercial activator, and control on the profile of the volatile compound families of Palomino Fino, Riesling, and Tintilla de Rota wines, respectively. It can be observed that bee pollen use significantly increased the total content of volatile compounds, resulting in an increment over the control by 3, 4, and 12% in the Palomino Fino (Table 1), Riesling (Table 2) and Tintilla de Rota (Table 3) wines, respectively ($p < 0.05$, ANOVA). However, a marked decrease in the production of volatile compounds in Palomino Fino and Riesling wines, around 11 and 22%, respectively, was observed with the commercial activator. In Tintilla de Rota wines, the values remained at the same levels as control (Table 3).

As can be seen, the higher alcohol levels in both white and red wines reached higher values using bee pollen, in comparison with the control and the synthetic activator, while the volatile acid content decreased in all wines with pollen, possibly due to the richness of bee pollen in fatty acids [15]. This decrease in volatile acids with pollen use favors the aromatic profile of the wines, since most of the compounds of this family have aromas within the fatty and rancid series, which can negatively affect the aromatic quality of the wines. Regarding C6 alcohols, the results showed an increase in Palomino Fino wines with pollen, in comparison with the control. In Riesling and Tintilla de Rota wines, significant reductions were observed with the use of pollen and commercial activators. In addition, alcohols decreased in all pollen and commercial activator wines compared to their controls. Considering terpenes and esters, all the bee pollen wines experienced a significant increase in their content, reaching much higher values in the case of esters. With respect to aldehydes, an increase in this family of compounds was only observed in Riesling white wines, noting the very low levels reached for both varieties. Lastly, phenols, only detected in red wines, showed an opposite behavior between pollen and synthetic activator. In red pollen wines, over 26% increase was observed in comparison with the values obtained in controls and wines with commercial activator.

Table 1. Volatile compound concentration (μg/L) in Palomino Fino white wines (control, pollen, and commercial activator).

Volatile Compound	Control	Pollen	Commercial
Higher alcohols			
2-Propanol	419,790.9 ± 7150.0 [a]	400,774.7 ± 14,870.0 [a]	404,816.4 ± 2700.0 [a]
1-Propanol	7851.7 ± 80.0 [a]	4734.1 ± 100.0 [b]	4864.3 ± 60.0 [b]
2-Methyl-1-propanol	29,588.9 ± 1040.0 [a]	29,268.0 ± 820.0 [a]	32,853.1 ± 770.0 [b]
3-Methyl-1-butanol	268,385.1 ± 7622.0 [a]	347,966.3 ± 915.2 [b]	229,924.9 ± 4033.7 [c]
Total	725,616.6 ± 15,892.0	782,743.0 ± 16,705.2	672,458.7 ± 7563.7
% higher alcohols	86.16%	90.20%	90.22%
Methanol	38,822.5 ± 1480.0 [a]	38,881.3 ± 1690.0 [a]	38,432.2 ± 720.0 [a]
Total	38,822.5 ± 1480.0	38,881.3 ± 1690.0	38,432.2 ± 720.0
% methanol	4.61%	4.48%	5.16%
Acids			
Butanoic acid	32.8 ± 0.7 [a]	30.2 ± 0.1 [a]	31.0 ± 0.3 [a]
3-Methyl-butanoic acid	200.0 ± 0.3 [a]	199.8 ± 1.8 [a]	220.5 ± 1.0 [b]
Hexanoic acid	1811.4 ± 176.4 [a]	627.2 ± 3.0 [b]	964.9 ± 18.3 [c]
Heptanoic acid	37.6 ± 1.5 [a]	38.4 ± 2.1 [a]	75.2 ± 3.4 [b]
2-Hexenoic acid	37.2 ± 0.4 [a]	42.0 ± 0.8 [b]	45.2 ± 0.6 [b]
Octanoic acid	2790.4 ± 131.3 [a]	1427.5 ± 4.3 [b]	1899.9 ± 33.6 [c]
Nonanoic acid	15.3 ± 1.2 [a]	6.3 ± 0.3 [b]	5.0 ± 0.1 [b]
n-Decanoic acid	794.9 ± 14.0 [a]	638.0 ± 35.8 [b]	926.1 ± 7.5 [c]
9-Decenoic acid	248.6 ± 32.3 [a]	108.1 ± 1.4 [b]	164.1 ± 8.9 [c]
Benzoic acid	94.1 ± 1.0 [a]	80.6 ± 4.7 [b]	101.2 ± 12.6 [a]
Phenylacetic acid	9.8 ± 0.3 [a]	15.0 ± 0.7 [b]	8.4 ± 0.6 [c]
Total	6072.0 ± 359.4	3213.1 ± 54.9	4441.5 ± 86.8
% acids	0.72%	0.37%	0.60%
C-6 Alcohols			
1-Hexanol	752.0 ± 7.8 [a]	590.2 ± 4.3 [a]	613.3 ± 9.0 [a]
(E)-3-hexen-1-ol	15.2 ± 0.3 [a]	21.1 ± 1.5 [a]	16.2 ± 1.0 [a]
(Z)-3-hexen-1-ol	56.2 ± 64.2 [a]	105.4 ± 1.7 [b]	119.4 ± 3.6 [b]
Total	823.5 ± 72.3	716.6 ± 7.4	748.9 ± 13.5
% C-6 alcohols	0.10%	0.08%	0.10%
Alcohols			
3-Penten-2-ol	10.0 ± 0.5 [a]	18.1 ± 0.2 [b]	15.2 ± 2.0 [c]
1-Pentanol	2185.7 ± 7.0 [a]	1474.5 ± 102.0 [b]	1170.5 ± 12.4 [c]
3-Ethyl-2-pentanol	8.2 ± 0.1 [a]	10.2 ± 0.2 [b]	7.2 ± 0.2 [c]
4-Methyl-1-pentanol	10.9 ± 0.6 [a]	15.7 ± 0.1 [b]	16.7 ± 0.1 [b]
3-Methyl-1-pentanol	92.8 ± 5.6 [a]	114.7 ± 0.3 [b]	117.6 ± 5.6 [b]
3-Ethoxy-1-propanol	87.7 ± 0.7 [a]	79.5 ± 0.7 [a,b]	76.8 ± 0.4 [b]
1-Octanol	17.1 ± 1.9 [a]	43.2 ± 1.0 [b]	43.0 ± 2.5 [b]
1-Nonanol	8.9 ± 0.1 [a]	11.2 ± 1.5 [b]	12.0 ± 0.7 [b]
Benzyl alcohol	44.8 ± 1.2 [a]	78.9 ± 1.6 [b]	41.6 ± 1.0 [a]
2-Phenylethanol	1236.5 ± 316.7 [a]	2119.0 ± 111.5 [b]	2102.9 ± 6.0 [b]
1H-Indole-3-ethanol	3240.6 ± 113.5 [a]	132.9 ± 0.1 [b]	105.0 ± 3.9 [b]
Total	6943.3 ± 447.9	4098.0 ± 219.2	3708.5 ± 34.9
% alcohols	0.82%	0.47%	0.50%
Terpenes			
Linalool oxide	7.0 ± 0.2 [a]	10.9 ± 0.9 [b]	10.1 ± 0.1 [b]
Linalool	8.1 ± 0.2 [a]	15.1 ± 0.9 [b]	10.8 ± 0.5 [c]
α-Terpineol	11.0 ± 0.3 [a]	15.8 ± 0.5 [b]	14.0 ± 0.7 [c]
β-Citronellol	9.4 ± 0.7 [a]	13.8 ± 0.8 [b]	10.1 ± 0.1 [a]
2,6-Dimethyl-3,7-octadiene-2,6-diol	29.3 ± 2.5 [a]	32.3 ± 1.5 [a]	22.3 ± 1.3 [b]
8-Hydroxilinalool	40.9 ± 0.1 [a]	48.9 ± 4.2 [b]	39.4 ± 3.0 [a]
Total	105.6 ± 3.9	136.8 ± 8.8	106.6 ± 5.7
% terpenes	0.013%	0.016%	0.014%

Table 1. *Cont.*

Volatile Compound	Control	Pollen	Commercial
Esters			
Ethyl acetate	11,164.9 ± 90.0 [a]	11,469.9 ± 20.0 [a]	8783.8 ± 350.0 [b]
Ethyl butyrate	5.8 ± 0.7 [a]	56.1 ± 2.1 [b]	36.3 ± 1.3 [c]
Ethyl isovalerate	7.1 ± 0.4 [a]	19.1 ± 0.0 [b]	8.3 ± 0.1 [a]
Isoamyl acetate	65.6 ± 0.8 [a]	79.3 ± 1.2 [b]	52.9 ± 0.8 [c]
Ethyl hexanoate	114.9 ± 5.1 [a]	173.7 ± 3.0 [b]	119.8 ± 14.0 [a]
Hexyl acetate	2.8 ± 0.2 [a]	4.3 ± 0.9 [b]	1.8 ± 0.2 [c]
Ethyl 2-hydrody-3-methyl butanoate	2.9 ± 0.1 [a]	10.7 ± 0.6 [b]	8.9 ± 0.1 [c]
Ethyl octanoate	215.5 ± 18.4 [a]	436.9 ± 37.5 [b]	207.4 ± 6.6 [a]
Ethyl nonanoate	8.3 ± 0.8 [a]	15.7 ± 0.2 [b]	12.6 ± 0.6 [c]
Ethyl 2-hydroxy-4-methylpentanoate	14.7 ± 2.9 [a]	26.5 ± 0.1 [b]	22.0 ± 1.8 [c]
Isoamyl lactate	14.6 ± 0.6 [a]	23.6 ± 0.4 [b]	15.9 ± 0.1 [a]
Ethyl decanoate	46.6 ± 1.1 [a]	57.5 ± 2.1 [b]	48.0 ± 0.2 [a]
Diethyl succinate	572.6 ± 18.4 [a]	954.7 ± 57.7 [b]	460.4 ± 31.8 [c]
Ethyl 9-decenoate	61.2 ± 1.5 [a]	80.0 ± 1.6 [b]	61.4 ± 1.6 [a]
Ethyl phenylacetate	1.2 ± 0.1 [a]	3.4 ± 0.2 [b]	2.3 ± 0.2 [c]
Phenethyl acetate	54.3 ± 5.0 [a]	108.8 ± 1.3 [b]	97.5 ± 3.4 [c]
Diethyl malate	28.3 ± 0.8 [a]	47.7 ± 1.6 [b]	29.6 ± 3.5 [a]
Ethyl 3-hydroxytridecanoate	28.7 ± 0.1 [a]	91.5 ± 2.8 [b]	66.1 ± 4.6 [c]
Methyl vanillate	1.7 ± 0.1 [a]	6.2 ± 0.0 [b]	4.3 ± 0.3 [c]
Total	1246.6 ± 146.9	13,665.6 ± 133.5	10,039.3 ± 421.2
% esters	0.15%	1.58%	1.35%
Aldehydes			
Acetaldehyde	62,513.36 ± 3140.0 [a]	24,201.6 ± 980.0 [b]	15,325.5 ± 200.0 [c]
Benzeneacetaldehyde	77.0 ± 0.7 [a]	103.3 ± 1.4 [b]	68.9 ± 1.4 [c]
Total	62,590.3 ± 3140.7	24,304.9 ± 981.4	15,394.4 ± 201.4
% aldehydes	7.43%	2.80%	2.07%

Different letters in superscript mean significant differences ($p < 0.05$) from two-way ANOVA on the basis of Bonferroni's multivariate test (BSD).

Table 2. Volatile compound concentration (μg/L) in Riesling white wines (control, pollen, and commercial activator).

Volatile Compound	Control	Pollen	Commercial
Higher alcohols			
2-Propanol	475,417.7 ± 5480.0 [a]	489,359.4 ± 23,570.0 [a]	297,882.7 ± 14,560.0 [b]
1-Propanol	2090.0 ± 50.0 [a]	1580.0 ± 60.0 [b]	2090.0 ± 60.0 [a]
2-Methyl-1-propanol	13,315.3 ± 223.3 [a]	11,882.7 ± 580.0 [b]	13,171.6 ± 236.7 [a]
3-Methyl-1-butanol	151,227.3 ± 6537.1 [a]	152,178.0 ± 7425.3 [a,b]	163,143.3 ± 3056.3 [b]
Total	642,050.2 ± 12290.4	655,000.2 ± 31,635.3	476,287.6 ± 17,913.0
% higher alcohols	91.64%	89.32%	87.46%
Methanol	19,623.5 ± 920.0 [a]	15,773.9 ± 410.0 [b]	15,497.5 ± 760.0 [b]
Total	19,623.5 ± 920.0	15,773.9 ± 410.0	15,497.5 ± 760.0
% methanol	2.80%	2.15%	2.85%

Table 2. *Cont.*

Volatile Compound	Control	Pollen	Commercial
Acids			
Butanoic acid	9.6 ± 0.1 [a]	20.7 ± 0.7 [b]	26.0 ± 0.7 [c]
3-Methyl-butanoic acid	94.3 ± 4.3 [a]	113.1 ± 0.9 [b]	127.0 ± 3.6 [b]
Hexanoic acid	1397.3 ± 99.7 [a]	1407.5 ± 222.8 [a]	1460.2 ± 181.3 [a]
Heptanoic acid	22.3 ± 2.3 [a]	22.7 ± 2.8 [a]	36.2 ± 6.0 [b]
2-Hexenoic acid	33.6 ± 0.7 [a]	55.7 ± 0.3 [b]	56.8 ± 0.3 [b]
Octanoic acid	2746.1 ± 153.7 [a]	1242.1 ± 1.1 [b]	2928.1 ± 233.2 [a]
Nonanoic acid	7.2 ± 0.3 [a]	15.7 ± 0.1 [b]	19.7 ± 0.8 [c]
n-Decanoic acid	96.1 ± 2.4 [a]	197.2 ± 1.5 [b]	194.1 ± 1.1 [b]
9-Decenoic acid	89.1 ± 6.7 [a]	152.9 ± 7.2 [b]	175.5 ± 4.6 [c]
Benzoic acid	77.7 ± 2.3 [a]	122.0 ± 0.3 [b]	125.8 ± 0.1 [b]
Phenylacetic acid	8.1 ± 0.2 [a]	16.0 ± 0.7 [b]	17.1 ± 0.4 [b]
Total	4581.4 ± 272.6	3365.5 ± 238.4	5166.5 ± 432.0
% acids	0.65%	0.46%	0.95%
C-6 Alcohols			
1-Hexanol	432.1 ± 10.6 [a]	554.3 ± 10.8 [b]	459.5 ± 3.6 [c]
(E)-3-hexen-1-ol	15.0 ± 0.6 [a]	37.6 ± 0.7 [b]	29.2 ± 0.7 [c]
(Z)-3-hexen-1-ol	26.7 ± 2.8 [a]	106.1 ± 2.5 [b]	85.0 ± 4.0 [c]
Total	473.8 ± 14.0	697.9 ± 14.0	573.6 ± 8.4
% C-6 alcohols	0.07%	0.10%	0.11%
Alcohols			
3-Penten-2-ol	13.5 ± 1.4 [a]	23.9 ± 0.7 [b]	27.9 ± 2.4 [c]
1-Pentanol	1190.6 ± 81.9 [a]	1094.6 ± 36.0 [b]	1245.0 ± 57.6 [a]
3-Ethyl-2-pentanol	6.3 ± 0.1 [a]	22.4 ± 0.4 [b]	24.3 ± 1.7 [c]
4-Methyl-1-pentanol	13.7 ± 0.8 [a]	33.3 ± 2.2 [b]	43.0 ± 1.6 [c]
3-Methyl-1-pentanol	62.6 ± 4.2 [a]	156.4 ± 2.0 [b]	89.3 ± 8.8 [c]
3-Ethoxy-1-propanol	8.5 ± 0.1 [a]	9.6 ± 0.2 [b]	10.7 ± 0.8 [c]
1-Octanol	36.9 ± 1.4 [a]	42.5 ± 0.0 [b]	30.8 ± 1.0 [c]
1-Nonanol	8.9 ± 0.3 [a]	27.0 ± 0.3 [b]	24.6 ± 0.1 [c]
Benzyl alcohol	54.2 ± 0.6 [a]	64.8 ± 1.9 [b]	43.5 ± 1.5 [c]
2-Phenylethanol	4432.3 ± 189.3 [a]	1200.7 ± 114.5 [b]	1047.5 ± 14.1 [b]
1H-Indole-3-ethanol	617.2 ± 0.3 [a]	116.6 ± 2.4 [b]	104.2 ± 1.4 [b]
Total	6444.6 ± 280.3	2791.7 ± 160.5	2690.6 ± 91.0
% alcohols	0.92%	0.38%	0.49%
Terpenes			
Linalool oxide	9.0 ± 0.2 [a]	16.3 ± 0.8 [b]	11.7 ± 0.7 [c]
Linalool	4.5 ± 0.3 [a]	15.1 ± 1.5 [b]	11.1 ± 0.2 [c]
α-Terpineol	12.2 ± 0.7 [a]	26.8 ± 0.1 [b]	17.2 ± 0.7 [c]
β-Citronellol	4.5 ± 0.2 [a]	16.5 ± 1.1 [b]	16.0 ± 0.3 [b]
2,6-Dimethyl-3,7-octadiene-2,6-diol	70.4 ± 0.1 [a]	94.6 ± 1.7 [b]	66.3 ± 4.5 [a]
8-Hydroxilinalool	6.7 ± 0.4 [a]	24.2 ± 1.6 [b]	22.5 ± 2.0 [b]
Total	107.4 ± 1.9	193.4 ± 6.8	144.8 ± 8.4
% terpenes	0.02%	0.03%	0.03%

Table 2. *Cont.*

Volatile Compound	Control	Pollen	Commercial
Esters			
Ethyl acetate	17,797.2 ± 590.0 [a]	12,172.1 ± 314.1[b]	13,628.0 ± 80.0 [c]
Ethyl butyrate	7.3 ± 0.1 [a]	25.8 ± 0.2 [b]	21.3 ± 1.2 [c]
Ethyl isovalerate	2.1 ± 0.1 [a]	10.9 ± 0.5 [b]	9.7 ± 0.4 [c]
Isoamyl acetate	111.2 ± 11.4 [a]	589.9 ± 4.5 [b]	560.6 ± 53.0 [b]
Ethyl hexanoate	51.0 ± 0.6 [a]	244.2 ± 3.1 [b]	159.5 ± 11.5 [c]
Hexyl acetate	2.7 ± 0.2 [a]	13.6 ± 0.8 [b]	13.2 ± 0.3 [b]
Ethyl 2-hydrody-3-methyl butanoate	4.2 ± 0.4 [a]	16.1 ± 0.4 [b]	12.6 ± 1.0 [c]
Ethyl octanoate	565.6 ± 13.4 [a]	757.3 ± 55.7 [b]	410.1 ± 5.4 [c]
Ethyl nonanoate	9.0 ± 0.2 [a]	17.0 ± 0.6 [b]	14.0 ± 0.2 [c]
Ethyl 2-hydroxy-4-methylpentanoate	19.8 ± 1.6 [a]	60.0 ± 4.4 [b]	58.9 ± 5.7 [b]
Isoamyl lactate	21.2 ± 0.7 [a]	43.0 ± 1.2 [b]	21.7 ± 2.1 [a]
Ethyl decanoate	55.0 ± 2.6 [a]	124.9 ± 10.9 [b]	121.6 ± 3.5 [b]
Diethyl succinate	663.0 ± 0.1 [a]	1435.3 ± 63.4 [b]	871.8 ± 83.3 [c]
Ethyl 9-decenoate	13.1 ± 0.3 [a]	32.3 ± 0.1 [b]	29.3 ± 0.8 [c]
Ethyl phenylacetate	2.0 ± 0.2 [a]	4.1 ± 0.1 [b]	2.4 ± 0.2 [c]
Phenethyl acetate	58.7 ± 1.0 [a]	242.7 ± 14.4 [b]	170.6 ± 8.7 [c]
Diethyl malate	30.8 ± 0.7 [a]	60.7 ± 0.6 [b]	40.4 ± 0.8 [c]
Ethyl 3-hydroxytridecanoate	10.5 ± 0.1 [a]	24.1 ± 1.2 [b]	16.7 ± 1.4 [c]
Methyl vanillate	18.6 ± 0.3 [a]	46.8 ± 0.6 [b]	36.3 ± 4.2 [c]
Total	1645.8 ± 624.0	15,920.5 ± 476.8	16,198.7 ± 263.5
% esters	0.23%	2.17%	2.97%
Aldehydes			
Acetaldehyde	25,673.4 ± 13,60.0 [a]	39,435.7 ± 1410.0 [b]	27,900.3 ± 1280.0 [a]
Benzeneacetaldehyde	30.0 ± 0.1 [a]	101.8 ± 1.0 [b]	97.8 ± 1.6 [b]
Total	25,703.4 ± 1360.1	39,537.5 ± 1411.0	27,998.1 ± 1281.6
% aldehydes	3.67%	5.39%	5.14%

Different letters in superscript mean significant differences ($p < 0.05$) from two-way ANOVA on the basis of Bonferroni's multivariate test (BSD).

Table 3. Volatile compound concentration (μg/L) in Tintilla de Rota red wines (control, pollen, and commercial activator).

Volatile compound	Control	Pollen	Commercial
Higher alcohols			
2-Propanol	293,773.7 ± 13,302.9 [a]	379,849.2 ± 14,140.0 [b]	302,480.2 ± 8707.4 [a]
1-Propanol	1306.8 ± 20.0 [a]	4529.0 ± 20.0 [b]	3141.0 ± 60.0 [c]
2-Methyl-1-propanol	23,210.1 ± 500.0 [a]	19,253.6 ± 760.0 [b]	20,531.5 ± 820.0 [b]
3-Methyl-1-butanol	246,833.3 ± 3329.3 [a]	276,725.2 ± 6681.7 [b]	273,487.4 ± 10,275.4 [b]
Total	565,123.9 ± 17,152.3	680,357.1 ± 21601.7	599,640.2 ± 19,862.8
% higher alcohols	80.20%	86.04%	84.34%
Methanol	63,489.4 ± 3570.0 [a]	55,928.6 ± 1160.0 [b]	61,709.1 ± 1139.8 [a]
Total	63,489.4 ± 3570.0	55,928.6 ± 1160.0	61,709.1 ± 1139.8
% methanol	9.01%	7.07%	8.68%

Table 3. *Cont.*

Volatile compound	Control	Pollen	Commercial
Acids			
Butanoic acid	29.4 ± 0.7 [a]	30.7 ± 0.7 [a]	33.8 ± 0.4 [b]
3-Methyl-butanoic acid	190.4 ± 2.6 [a]	177.2 ± 2.4 [a]	181.2 ± 1.4 [a]
Hexanoic acid	572.4 ± 24.4 [a]	446.0 ± 59.8 [b]	370.9 ± 18.2 [c]
Heptanoic acid	26.0 ± 0.9 [a]	23.3 ± 0.7 [b]	27.2 ± 0.9 [a]
2-Hexenoic acid	34.1 ± 1.2 [a]	38.4 ± 0.7 [b]	40.1 ± 0.1 [b]
Octanoic acid	1476.8 ± 93.5 [a]	883.8 ± 68.8 [b]	1185.9 ± 19.7 [c]
Nonanoic acid	88.5 ± 0.1 [a]	87.4 ± 2.5 [a]	88.3 ± 2.5 [a]
n-Decanoic acid	726.9 ± 7.4 [a]	362.2 ± 32.0 [b]	403.7 ± 15.4 [b]
9-Decenoic acid	36.2 ± 1.9 [a]	41.4 ± 0.7 [b]	43.3 ± 2.7 [b]
Benzoic acid	71.6 ± 1.4 [a]	130.5 ± 14.1 [b]	134.6 ± 15.2 [b]
Phenylacetic acid	46.1 ± 3.6 [a]	39.6 ± 2.3 [b]	52.8 ± 0.6 [c]
Total	3298.4 ± 137.6	2260.6 ± 184.7	2561.8 ± 77.0
% acids	0.47%	0.29%	0.36%
C-6 Alcohols			
1-Hexanol	405.8 ± 3.1 [a]	367.5 ± 34.7 [a]	325.6 ± 16.5 [b]
(E)-3-Hexen-1-ol	15.9 ± 0.4 [a]	31.1 ± 1.6 [b]	27.0 ± 0.7 [c]
(Z)-3-Hexen-1-ol	23.9 ± 0.8 [a]	26.7 ± 1.2 [b]	22.4 ± 1.4 [a]
Total	445.5 ± 4.3	425.3 ± 37.5	375.0 ± 18.5
% C-6-alcohols	0.06%	0.05%	0.05%
Alcohols			
3-Penten-2-ol	14.1 ± 1.1 [a]	27.9 ± 0.4 [b]	32.3 ± 1.8 [c]
1-Pentanol	1492.9 ± 119.4 [a]	1338.7 ± 28.2 [b]	1378.0 ± 35.7 [a,b]
3-Ethyl-2-pentanol	11.9 ± 2.1 [a]	13.1 ± 1.0 [a]	18.4 ± 1.4 [b]
4-Methyl-1-pentanol	15.6 ± 0.5 [a]	15.8 ± 0.2 [a]	21.5 ± 0.5 [b]
3-Methyl-1-pentanol	266.1 ± 23.0 [a]	130.3 ± 2.4 [b]	83.6 ± 2.9 [c]
3-Ethoxy-1-propanol	87.0 ± 0.9 [a]	123.3 ± 2.5 [b]	111.4 ± 1.9 [c]
1-Octanol	24.4 ± 1.7 [a]	25.1 ± 0.5 [a]	28.8 ± 0.9 [b]
1-Nonanol	3.7 ± 0.2 [a]	2.2 ± 0.2 [b]	1.9 ± 0.1 [b]
Benzyl alcohol	80.7 ± 0.8 [a]	140.4 ± 7.5 [b]	103.5 ± 11.6 [c]
2-Phenylethanol	2440.2 ± 6.0 [a]	2597.6 ± 133.5 [a]	2128.8 ± 6.6 [b]
1H-Indole-3-ethanol	854.8 ± 4.2 [a]	662.0 ± 40.6 [b]	673.8 ± 50.8 [b]
1-Butanol	32.7 ± 3.0 [a]	18.5 ± 0.9 [b]	10.4 ± 0.4 [c]
3-Methyl-2-buten-1-ol	47.0 ± 1.1 [a]	34.3 ± 2.2 [b]	32.7 ± 1.7 [b]
Total	5371.2 ± 164.1	5129.2 ± 220.2	4625.2 ± 116.2
% alcohols	0.76%	0.65%	0.65%
Phenols			
2,6-Di-tert-butyl-4-ethylphenol	17.9 ± 0.1 [a]	30.6 ± 1.2 [b]	27.3 ± 3.3 [c]
4-Ethylphenol	5.8 ± 0.1 [a]	6.4 ± 0.1 [a]	5.8 ± 0.1 [a]
4-Vinylguaiacol	36.8 ± 1.7 [a]	64.2 ± 1.5 [b]	53.1 ± 0.6 [c]
Acetovanillone	89.3 ± 2.1 [a]	87.9 ± 5.5 [a]	61.2 ± 6.9 [b]
Total	149.9 ± 4.0	189.1 ± 8.3	147.4 ± 10.8
% phenols	0.02%	0.02%	0.02%
Terpenes and derivatives			
Linalool oxide	23.3 ± 1.6 [a]	65.4 ± 0.5 [b]	60.3 ± 4.4 [b]
Linalool	7.9 ± 0.6 [a]	14.3 ± 0.1 [b]	15.1 ± 1.3 [b]
α-Terpieol	8.9 ± 0.4 [a]	15.3 ± 0.9 [b]	12.8 ± 0.2 [c]
β-Citronellol	8.2 ± 0.1 [a]	10.8 ± 0.1 [b]	10.3 ± 0.7 [b]
2,6-Dimetil-3,7-octadiene-2,6-diol	30.5 ± 2.0 [a]	45.6 ± 1.5 [b]	39.9 ± 3.5 [c]
8-Hydroxylinalool	50.4 ± 3.5 [a]	194.1 ± 5.6 [b]	92.2 ± 3.0 [c]
Total	129.1 ± 8.2	345.6 ± 8.7	230.6 ± 13.0
Terpenes and derivatives	0.02%	0.04%	0.03%

Table 3. *Cont.*

Volatile compound	Control	Pollen	Commercial
Esters			
Ethyl acetate	34,390.6 ± 1370.0 [a]	25,802.0 ± 590.0 [b]	24,869.0 ± 530.0 [b]
Ethyl butyrate	24.5 ± 1.3 [a]	72.8 ± 1.3 [b]	54.5 ± 4.0 [c]
Ethyl isovalerate	27.4 ± 0.5 [a]	30.2 ± 1.9 [b]	15.7 ± 0.4 [c]
Isoamyl acetate	67.1 ± 3.3 [a]	76.2 ± 1.1 [b]	30.7 ± 1.0 [c]
Ethyl hexanoate	79.5 ± 2.4 [a]	157.9 ± 12.9 [b]	115.6 ± 4.2 [c]
Hexyl acetate	35.4 ± 1.8 [a]	76.6 ± 0.9 [b]	62.7 ± 1.0 [c]
Butanoic acid, 2-hydroxy-3-methyl-, ethyl ester	5.7 ± 0.3 [a]	16.8 ± 0.3 [b]	13.9 ± 0.8 [c]
Ethyl octanoate	333.1 ± 12.2 [a]	415.0 ± 32.5 [b]	317.2 ± 22.0 [a]
Ethyl nonanoate	10.1 ± 0.1 [a]	11.5 ± 0.4 [b]	10.4 ± 0.1 [a]
Ethyl 2-hydroxy-4-methylpentanoate	32.4 ± 2.3 [a]	43.0 ± 3.5 [b]	49.3 ± 0.1 [c]
Isoamyl lactate	48.0 ± 2.1 [a]	239.9 ± 4.4 [b]	208.0 ± 4.7 [c]
Ethyl decanoate	214.7 ± 5.6 [a]	278.6 ± 6.3 [b]	273.9 ± 16.3 [b]
Diethyl succinate	404.0 ± 17.0 [a]	971.2 ± 40.7 [b]	944.7 ± 66.2 [b]
Ethyl 9-decenoate	58.7 ± 0.6 [a]	63.9 ± 3.8 [b]	58.8 ± 0.2 [a]
Ethyl phenylacetate	1.2 ± 0.0 [a]	2.2 ± 0.0 [b]	2.4 ± 0.2 [c]
Phenethyl acetate	35.8 ± 1.4 [a]	140.4 ± 8.3 [b]	138.0 ± 5.7 [b]
Diethyl malate	22.4 ± 0.7 [a]	34.7 ± 1.6 [b]	30.0 ± 0.3 [c]
Methyl vanillate	147.7 ± 12.2 [a]	583.1 ± 8.8 [b]	404.3 ± 0.4 [c]
Ethyl lactate	121.6 ± 1.6 [a]	153.5 ± 1.3 [b]	151.0 ± 3.7 [b]
Butanoic acid 3-hydroxy, ethyl ester	56.1 ± 0.3 [a]	89.3 ± 6.3 [b]	73.3 ± 4.8 [c]
Ethyl (Z)-4-decenoate	56.8 ± 2.1 [a]	220.3 ± 2.4 [b]	128.3 ± 7.4 [c]
Ethyl dodecanoate	57.1 ± 1.0 [a]	65.2 ± 0.1 [b]	61.3 ± 4.3 [a,b]
Methyl tetradecanoate	36.2 ± 0.8 [a]	84.4 ± 6.4 [b]	66.8 ± 1.6 [c]
Succinic acid, 2-hydroxy-3-methyl-, diethyl ester	88.7 ± 1.6 [a]	130.9 ± 1.9 [b]	122.2 ± 2.9 [b]
Methyl hexadecanoate	77.2 ± 2.0 [a]	154.3 ± 1.0 [b]	122.8 ± 6.8 [c]
Hexadecanoic acid, ethyl ester	370.1 ± 25.1 [a]	754.2 ± 5.8 [b]	354.9 ± 10.5 [a]
Propanoic acid, 2-methyl, propyl, 2-methyl, ester	79.7 ± 0.6 [a]	166.0 ± 0.5 [b]	143.3 ± 4.8 [c]
Ethyl 8-nonenoate	201.6 ± 0.7 [a]	245.4 ± 2.9 [b]	220.4 ± 1.1 [c]
Total	2692.6 ± 1469.7	31,079.5 ± 747.4	29,043.3 ± 705.2
% esters	0.38%	3.93%	4.08%
Aldehydes			
Acetaldehyde	63,723.2 ± 2330.0 [a]	14,546.3 ± 458.5 [b]	12,209.5 ± 520.0 [b]
Benzeneacetaldehyde	63.6 ± 4.7 [a]	123.8 ± 2.2 [b]	120.6 ± 0.9 [b]
Nonanal	10.0 ± 0.2 [a]	16.5 ± 1.2 [b]	12.8 ± 0.6 [c]
3-Methyl-butanal	13.6 ± 1.6 [a]	30.1 ± 0.2 [b]	26.3 ± 1.1 [c]
Total	63,810.4 ± 2336.5	14,716.6 ± 462.1	12,369.2 ± 522.7
% aldehydes	9.06%	1.86%	1.74%
Thiols			
3-(Methylthio)-1-propanol	18.1 ± 1.6 [a]	121.7 ± 1.9 [b]	104.3 ± 3.3 [c]
Total	18.1 ± 1.6	121.7 ± 1.9	104.3 ± 3.3
% thiols	0.003%	0.02%	0.01%

Table 3. *Cont.*

Volatile compound	Control	Pollen	Commercial
Acetals			
1-(1-(1-Ethoxyethoxy)-pentane	1.9 ± 0.1 [a]	2.0 ± 0.1 [a]	2.1 ± 0.2 [a]
Total	1.9 ± 0.1	2.0 ± 0.1	2.1 ± 0.2
% acetals	0.0003%	0.0003%	0.0003%
Norisoprenoids			
3-Oxo-α-ionol	5.9 ± 0.1 [a]	12.2 ± 1.1 [b]	15.1 ± 1.0 [c]
Total	5.9 ± 0.1	12.2 ± 1.1	15.1 ± 1.0
% norisoprenoids	0.0008%	0.002%	0.002%
Lactones			
Dihydro-5-pentyl-2(3H)-furanone	56.1 ± 2.4 [a]	108.4 ± 0.4 [b]	99.7 ± 1.6 [c]
2,3-Dihydro-benzofuran	45.5 ± 1.8 [a]	45.8 ± 1.0 [a]	55.1 ± 0.2 [b]
Total	101.6 ± 4.2	154.2 ± 1.4	154.8 ± 1.8
% lactones	0.0144%	0.0219%	0.0220%

Different letters in superscript mean significant differences ($p < 0.05$) from two-way ANOVA on the basis of Bonferroni's multivariate test (BSD).

3.1.1. Major Alcohols and Methanol

The majority alcohol levels in wines (white and red) achieved greater values with the use of bee pollen compared to controls and commercial activators (Tables 1–3), especially Tintilla de Rota red wines (Table 3). In red wines, higher alcohols were significantly raised in similar proportions as in previous study [32], except for 2-methyl-1-propanol, which had the lowest concentration (Table 3). This behavior was also observed in white wines, mainly due to the increase of 3-methyl-1-butanol in Palomino Fino and 2-propanol and 3-methyl-1-butanol in Riesling wines [31]. Regarding methanol, we observed a slight increase in its concentration (0.15 %) in Palomino Fino pollen wines compared to the control, while the commercial activator wines decreased in methanol production (Table 1). The same behavior was observed in a previous research on different bee pollen doses in Palomino Fino wines [31]. On the other hand, in Riesling (Table 2) and Tintilla de Rota wines (Table 3), we observed a considerable decrease in methanol concentration, both with pollen and with the commercial activator, being more marked in red wines.

3.1.2. Volatile Acids

The volatile acid content was lower in all bee pollen wines (Tables 1–3) as a consequence of pollen fatty acid richness [15]. However, the behavior of wines with commercial activator differs for the white wines. In Palomino Fino wines, a decrease was observed (Table 1), while in Riesling wines (Table 2), there was a significant increase. In the case of Tintilla (Table 3), the reduction in acid content was mostly related to hexanoic, octanoic, and decanoic acids. The volatile acid family compounds decreased as a result of pollen use. This family of compounds disfavors the aromatic profile of the wines, as the majority volatile acids present aromas within the fatty and rancid series by altering the aromatic quality of wines [31,32].

3.1.3. C-6 Alcohols

In Palomino Fino wines (Table 1), the production of C6-alcohols decreased compared to the control by 47.08% with the use of pollen and 25.85% with the commercial activator, mainly due to 1-hexanol. On the contrary, in Riesling (Table 2), this family of compounds increased by 47.30% and 21.07% with pollen and commercial activator, respectively. In Tintilla de Rota wines (Table 3), the C6-alcohols levels decreased significantly using pollen, although with the commercial activator, there was a greater decrease compared to the control.

3.1.4. Alcohols and Terpenes

Alcohols decreased significantly compared to the control, both in pollen-elaborated wines and with commercial activator, highlighting the decrease in the commercial activator wines. The main responsible compounds for the decrease in this family in wines with pollen were 1-pentanol and 1H-indole-3-ethanol for all wines (Tables 1–3). In terms of the commercial activator, the main responsible compounds were 1-pentanol in Palomino Fino and 2-phenylethanol in Riesling wines. For Tintilla de Rota wines, the decrease in alcohols was caused by 1-pentanol and 1H-indole-3-ethanol. Regarding the terpene content, it was clearly observed that terpene levels were significantly higher in all pollen-elaborated wines, supporting pollen's ability to transfer this family of compounds to the wine.

3.1.5. Esters

As for esters, all wines showed a significant increase in their ester content, reaching a 10-fold increase over control wines. In Palomino Fino (Table 1) and Riesling (Table 2) wines, bee pollen produced more esters than the commercial activator, while for Tintilla de Rota wines, the values were similar to each other (Table 3). These results could demonstrate that pollen contributes significantly to the synthesis and formation of esters, intensifying fruity and floral aromatic series in wines.

3.1.6. Aldehydes

The levels of aldehydes reached for the white wines of both varieties were very low. Riesling wines showed an increase (Table 2), while Palomino Fino wines decreased with both bee pollen and commercial activator (Table 1). The same behavior was observed for Tintilla de Rota (Table 3) as for Palomino Fino white wines. Moreover, aldehyde levels were significantly lower in all pollen-elaborated and commercial activator-elaborated wines compared to the control, caused in particular by pronounced reductions in acetaldehyde. However, the levels of benzene-acetaldehyde, nonanal, and 3-methylbutanal were higher compared to the control wines.

3.1.7. Phenols and Minority Compounds

The phenol content, only detected in Tintilla de Rota wines, showed an opposite behavior between bee pollen and commercial activator wines. Pollen contributed more than 26% to the phenol levels of the control and the commercial activator wines. These compounds could possibly play a role in the spicy character of wines [41].

Among the families of minority compounds also detected in Tintilla de Rota, represented by thiols, acetals, and norisoprenoids, we observed that bee pollen and commercial activator addition enhanced thiol and norisoprenoid formation.

3.2. *Principal Component Analysis of Volatile Compounds (PCA)*

Table 4 shows the loadings for the principal component analysis (PCA) factors extracted on the dataset corresponding to the analyzed families of volatile compounds in the white and red wines: higher alcohols, methanol, acids, C6 alcohols, alcohols, phenols, terpenoids, esters, aldehydes, acetals, norisoprenoids, and lactones. Three factors were extracted from the data analysis, which explained over 88% variance data. Factor F1 was in positive correlation with most volatile minority compounds (thiols, methanol, acetals, norisoprenoids, lactones, and phenols) characteristic of Tintilla de Rota wines and negatively with C6 alcohols. Terpenes were also positively included in F1, with lower loadings. This effect was probably related to the influence of the grape skins on this group of compounds during Tintilla de Rota vinification. As can be seen in Figure 1a, all white wines showed negative F1 values, with the lowest values for the commercial activator and pollen wines. On the contrary, all red wines showed positive values (>1), highlighting bee pollen wines with the highest levels, supporting the enhancing influence on the minority compounds involved in Tintilla de Rota's varietal expression.

Factor 2 (F2) was positively correlated with the volatile compounds mostly responsible of the aromatic profile of wines (esters and terpenes) and negatively with aldehydes and alcohols. F2 therefore grouped the family of aromatic compounds most influenced by the presence of bee pollen. According to the results obtained, the use of bee pollen mainly promoted ester production and the transfer of terpenes, and decreased alcohol and aldehyde formation. Bee pollen was highly correlated with an increase in compounds responsible of aromatic notes of fruits and flowers, qualities determining during the sensory analysis of the wines. Another effect observed in F2 was a correlation with terpenes and derivatives, mainly as result of the synergistic effect of pollen and skins (red wines) upon this volatile compound family. Figure 1a,b shows the aroma-enhancing effect of pollen during vinification in white and red wine.

Table 4. Main component loads of volatile organic compounds in white and red wines (control, commercial activator, and bee pollen).

Volatile Compounds	F1	F2	F3
Higher alcohols	−0.097	0.009	0.986
Methanol	0.875	−0.155	0.195
Acids	−0.669	−0.535	−0.081
C-6 alcohols	−0.754	−0.144	0.476
Alcohols	0.356	−0.704	0.387
Terpenes and derivatives	0.668	0.614	0.068
Esters	0.497	0.834	−0.024
Aldehydes	−0.086	−0.832	−0.052
Thiols	0.873	0.418	0.059
Acetals	0.980	0.061	−0.148
Norisoprenoids	0.938	0.274	−0.062
Lactones	0.978	0.178	−0.089
Phenols	0.977	0.083	−0.108
Explained variance (%)	54.88	22.19	11.16

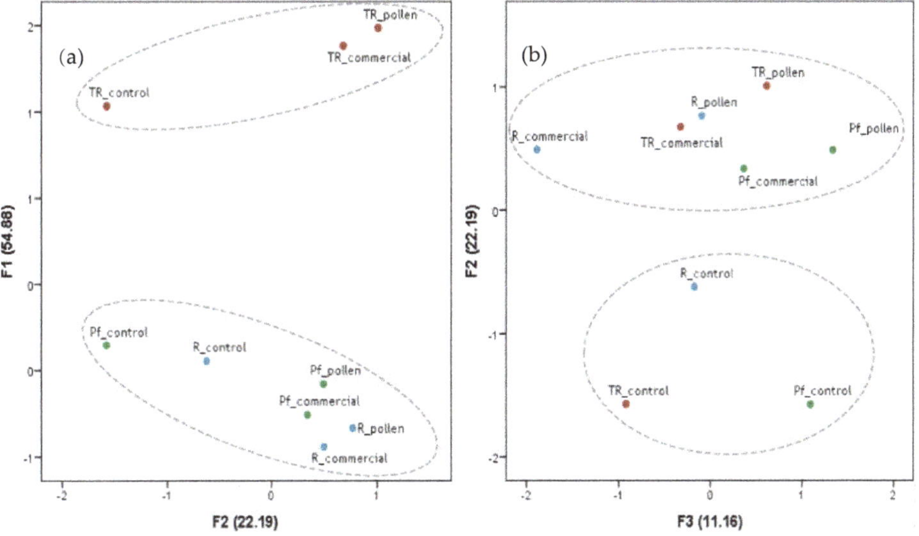

Figure 1. Principal component analysis of Palomino Fino (Pf, green), Riesling (R, blue), and Tintilla de Rota (TR, red) wines fermented with bee pollen against a commercial activator and control.

All wines with bee pollen and commercial activator showed positive values in F2, highlighting pollen-elaborated wines. This would indicate that the use of bee pollen in winemaking enhances, on one hand, the formation and, on the other hand, the extraction of compounds contributing to the aromas in wines.

The third factor (F3) mainly correlated and grouped higher alcohols, the majority volatiles in the wines. Pollen had a clear effect on the behavior of this compound family with respect to both wines, white and red, enhancing higher alcohol production. Figure 1b shows the two aromatic factors (F2 and F3) where bee pollen exerted a certain effect. As can be seen, all the wines with bee pollen had the highest F3 levels in their respective series compared to the control and to the commercial activator wines.

3.3. Description of Odor Activity Value (OAV)

Tables 5 and 6 show the values of the sum of the OAV of each aromatic series, as well as the sum of all the series (ΣOAV_T).

Table 5. Sum of the odorant activity values (ΣOAV) of Palomino Fino and Riesling white wines, grouped by odorant series.

	Palomino Fino			Riesling		
Odorant Series	**Control**	**Pollen**	**Commercial**	**Control**	**Pollen**	**Commercial**
Fruity	85.74	154.02	85.69	138.28	239.97	149.42
Floral	17.63	24.61	16.91	8.56	25.42	23.23
Fatty	17.09	11.28	14.01	11.83	9.59	13.57
Grassy	1.06	1.03	1.05	0.58	0.99	0.81
Dried fruit	6.24	2.43	1.54	2.56	3.93	2.78
Earthy, mushroom	0.30	0.28	0.29	0.15	0.31	0.19
Chemical	0.09	0.09	0.10	0.08	0.12	0.13
Spicy	0.03	0.05	0.04	0.01	0.02	0.05
ΣOAV_T	128.18	193.79	119.62	162.05	280.36	190.17

Table 6. Sum of odorant activity values (ΣOAV) of Tintilla de Rota red wines, grouped by odorant series.

Odorant Series	Control	Pollen	Commercial
Fruity	125.11	173.85	133.14
Floral	17.61	37.74	36.23
Fatty	11.00	8.76	9.36
Grassy	0.54	0.64	0.54
Dried fruit	6.38	1.45	1.22
Earthy, mushroom	0.27	0.26	0.19
Chemical	0.08	0.13	0.13
Spicy	1.06	1.88	1.53
Phenolic	0.01	0.01	0.01
ΣOAV_T	162.05	224.71	182.35

The sum of the OAV groups together the different aromatic series in which the different volatile compounds are associated with their perception threshold according to the literature. In general, all the wines with pollen showed the highest ΣOAV_T values, with markedly higher variances in pollen wines versus the control and commercial activator. This can be explained by the fact that the wines with pollen had the highest levels in floral and fruity odor series among all the wines studied.

The commercial activator also enhanced or intensified the fruity and floral series for Riesling and Tintilla de Rota wines, although to a minor extent compared to bee pollen. The exception was Palomino Fino wines, where the fruity and floral odorant series values in the wines with commercial activator were very similar to the control.

Figure 2 show the increment percentage of "positive" odorant series (fruity + floral) versus "negative" odorant series (fatty + herbaceous + raisin fruit (nuts)) for Palomino Fino (Figure 2a) and Riesling (Figure 2b) white wines and Tintilla de Rota red wines (Figure 2c).

The greatest contribution to the aromatic profile in all cases was made by the "positive" odorant series, whose values were above 93–95% in the wines with bee pollen, followed by the commercial activator, with values above 83–93%, and finally the control with 81–90%.

Therefore, from a general point of view, it could be pointed out that pollen managed to enhance the formation of volatile compounds with positive odoring effects (fruit and flowers), as opposed to volatile compounds with negative odoring effects (fatty).

3.4. Descriptive Sensory Analysis

A sensorial analysis of all the wines was carried out to describe the organoleptic attributes that described them in detail. A specific tasting sheet was used for the white wines and another sheet for the red wines. In order to carry out the comparative study in detail, we made a distinction between the general tasting attributes and the specific attributes for each type of wine.

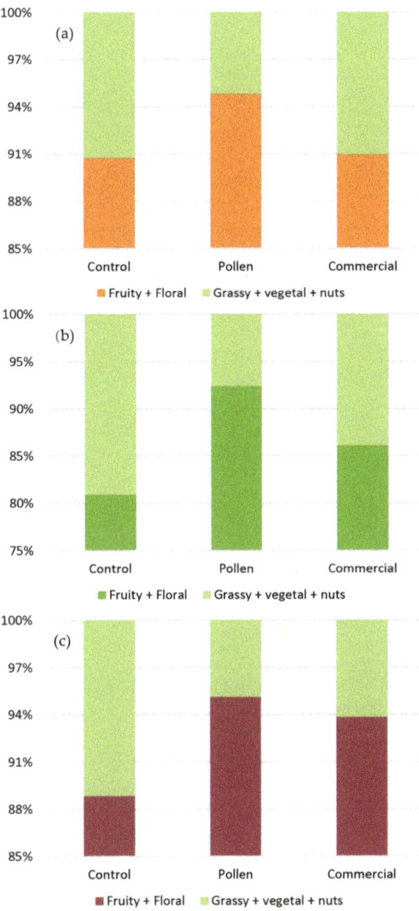

Figure 2. Participation percentage of the odorant series (fruity, floral, fatty, herbaceous, and dry fruit) in the white wines of Palomino Fino (**a**) and Riesling (**b**), and in Tintilla de Rota red wines (**c**).

3.4.1. General Attributes in White Wines

Figures 3 and 4 show the cobweb diagrams corresponding to the sensory profile of the general attributes for Palomino Fino and Riesling white wines, correspondingly (fruity, floral, and spicy character on the nose; sweetness, acidity, bitterness, astringency, salinity, body, and persistence). As can be seen, in both cases, wines with pollen scored significantly higher in the fruit and floral attributes related to the control and the commercial activator, whereby control wines obtained the lowest scores. These scores verified the OAV results and probably reflect that pollen specifically enhances the synthesis of esters during alcoholic fermentation.

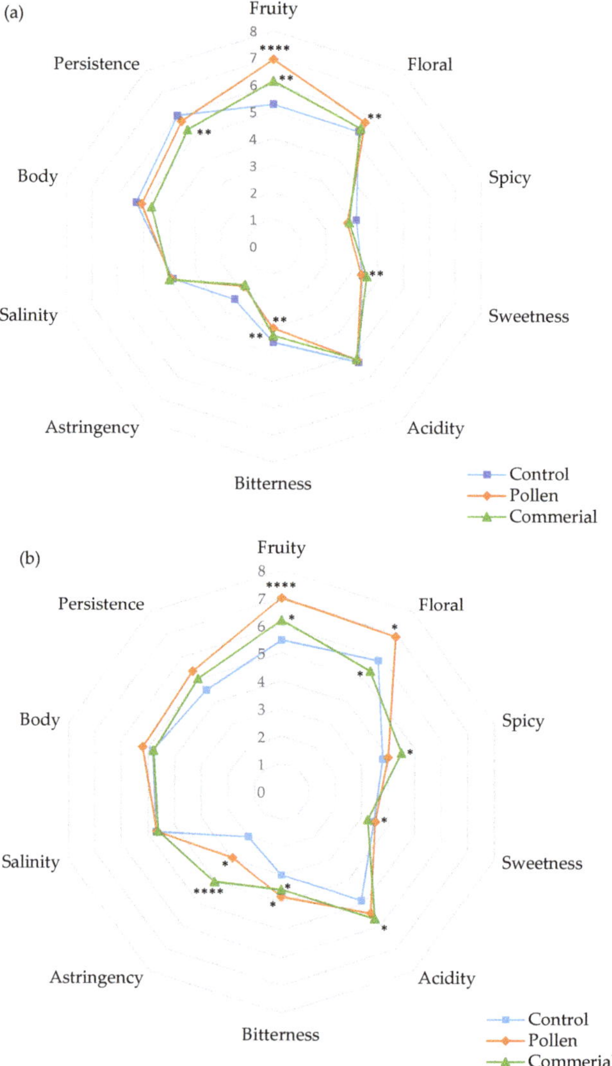

Figure 3. General attributes of the sensory analysis of Palomino Fino (**a**) and Riesling (**b**) wines (control, pollen, and commercial activator). * Statistical significance for two-way ANOVA (BSD test) (* $p < 0.05$, ** $p < 0.01$,, and **** $p < 0.0001$).

In Palomino wines, floral character scored no significant differences between control and commercial activator wines (Figure 3a). However, control wines showed the best scores compared to the commercial activator in Riesling wines (Figure 3b).

Regarding the sweet mouthfeel attribute, the control and pollen wines showed lower scores compared to the wines with commercial activator, while in Riesling, control and pollen wines scored slightly higher compared to the wines with commercial activator. In terms of acidity, all the samples showed moderate acidity levels, and only the commercial activator and pollen wines were slightly higher than the control. No wine stood out for its bitter character; however, the control in Palomino Fino scored significantly higher than pollen and commercial activator, while in Riesling, the behavior was the opposite. Astringency of these wines was rated very low in general, with only Riesling wines with commercial activator receiving the highest scores. As for the saline character, all the wines showed low values with no significant differences.

Concerning the texture sensation in the mouth, we should note that the wines with the most body were control and pollen in Palomino Fino, and pollen wines in Riesling. Synthetic activator wines obtained the lowest scores in both cases. Moreover, it was the same samples that presented significantly higher persistency in Palomino wines, while pollen and commercial activator Riesling wines were better scored.

3.4.2. Specific Attributes in White Wines

The specific attributes (white fruit, tropical fruit, citrus fruit, stone fruit, dried fruit, flowers, spicy aromas, balsamic, etc.) evaluated in the sensory analysis are shown in Figure 4. A detailed analysis of specific attributes of the olfactory sensory phase among white wines showed that, in general, Riesling (Figure 4a) presented a more complex and richer sensory profile than Palomino Fino wines (Figure 4b).

While in Palomino Fino wines, white fruit notes (apple and pear) were significantly intensified, in Riesling, in addition to white fruit, tropical fruit (banana, melon, pineapple, passion fruit) and citrus notes (tangerine and lime) were increased compared to the commercial activator and the control. Palomino Fino wines with commercial activator and pollen showed higher intensity in stone fruit and lower intensity in raisin fruit compared to the control, despite the latter attribute scoring very low in all cases. Only a slight increase in raisin fruit notes was observed in the Riesling wines with pollen.

In Palomino Fino pollen and commercial activator wines, the highest scores were attributed to the intensity of white flowers (orange blossom, jasmine), while the rest of the attributes were assessed by tasters with very low scores. The intensity of white flowers was higher in Riesling wines than in Palomino Fino, in line with the increases observed in this odor series in the OAV analysis. Riesling wines made with the commercial activator showed increases in balsamic, microbiological, and chemical notes, not observed in pollen wines. This behavior suggests although a more complex sensory profile was achieved with the Riesling variety, the intensity of these minority attributes seemed to be integrated in a well-balanced way. Regarding the warmth attribute, Riesling wines with pollen had a warmer mouthfeel, possibly related to the slight increase in body and persistence noted.

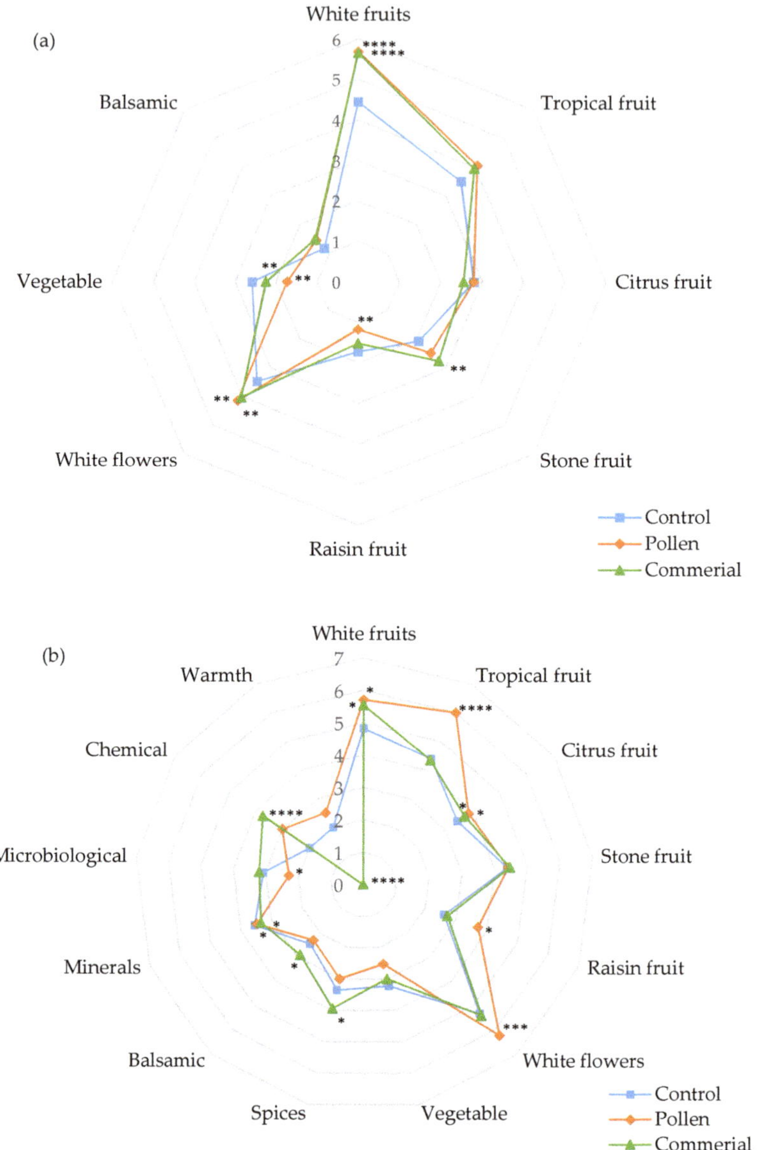

Figure 4. Specific attributes of sensory analysis of Palomino Fino (**a**) and Riesling (**b**) wines (control, pollen, and commercial activator). * Statistical significance for two-way ANOVA (BSD test) (* $p < 0.05$, ** $p < 0.01$, *** $p < 0.001$, and **** $p < 0.0001$).

3.4.3. General and Specific Attributes in Red Wines

Figure 5 represents the results obtained for the general attributes after the sensory analysis of the Tintilla de Rota red wines (control, pollen, and commercial).

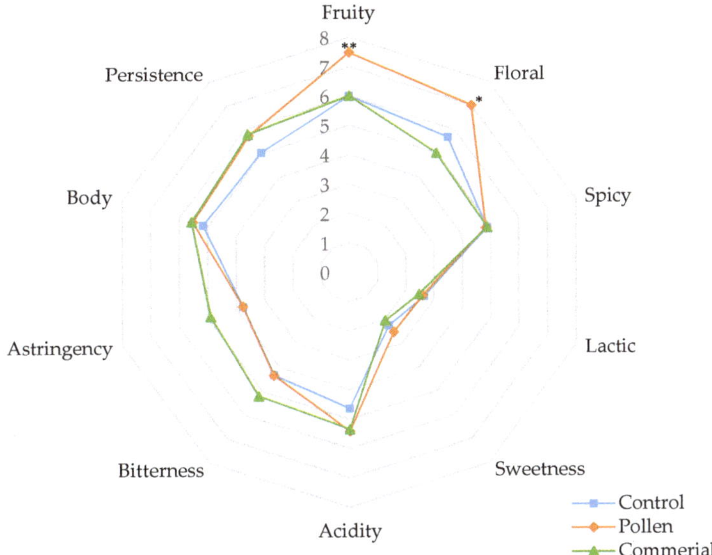

Figure 5. General attributes of the sensory analysis of Tintilla de Rota wines (control, pollen, and commercial). * Indicates level of significance for two-way ANOVA (BSD test) (* $p < 0.05$ and ** $p < 0.01$).

In general, Tintilla de Rota wines fermented with bee pollen presented significantly higher scores in fruity and floral attributes in comparison to the control and the commercial activator. No variation in sweetness intensity was observed in any of the cases. However, a higher score was observed in the acidity intensity in wines with bee pollen and commercial activator. Bitterness and astringency sensations together showed very different behavior in the wines. The wines with pollen and control were the ones with the lowest scores in both attributes. As with the Riesling wines, red wines with pollen and commercial activator were the most intense in the attributes of mouthfeel, body, and persistence.

Specific attributes evaluated in the sensory analysis of the Tintilla de Rota red wines (control, pollen, and commercial) are shown in Figure 6.

Considering the specific olfactory attributes, all red wines with bee pollen showed significantly more richness and diversity in the fruity profile (blackberries, raspberries, cherries, apple, as well as melon, mango, passion fruit, tangerine, and lime notes) (ANOVA $p < 0.05$) compared to control and commercial activator wines. No substantial differences were observed in stone fruit, ripe fruit, and raisin fruit aromas. In relation to the floral attributes, the wines with pollen scored best in their notes of blue flowers (violets and lilacs) and red flowers (roses), and to a lesser extent in white flowers, where the control stood out. Both types of wines were better scored than the commercial activator. All other attributes showed no significant differences between the wines.

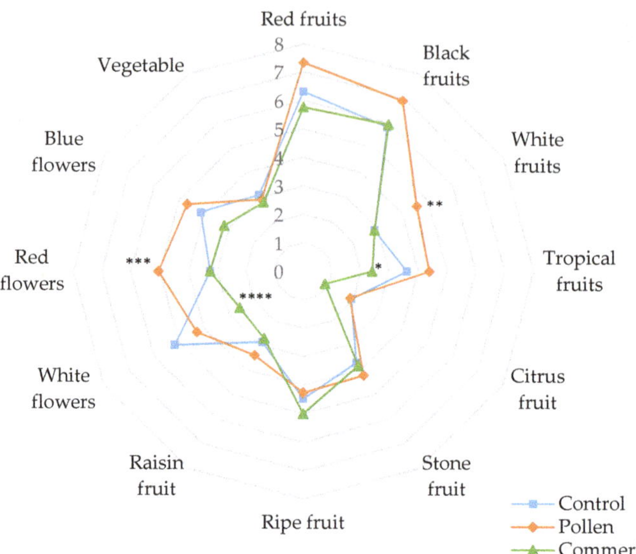

Figure 6. Specific attributes of the sensory analysis of Tintilla de Rota red wines (control, pollen, and commercial). * Statistical significance for two-way ANOVA (BSD test) (* $p < 0.05$,** $p < 0.01$, *** $p < 0.001$ and **** $p < 0.0001$).

3.5. Classification and Triangular Test

Once characterized by attributes (olfactory, taste, and texture), wines were tested by the classification test (ISO 8587:2006 AENOR, 2010) [38] and the triangular test (ISO 4120:2007 AENOR, 2008) [39].

Classification test results for white and red wines elaborated with pollen versus a commercial activator and control are shown in Table 7. Triangular test results for the obtained values of alpha risk (α), number of tasters, and percentage of tasters who identified the different wines in each category are shown in Table 8.

Table 7. Results of the classification test on white and red wines from the comparative study with pollen and commercial activator.

Variety	Control	Pollen	Commercial	Testers	F-Test	Significance (α)
Palomino Fino	42 (24.1%)	66 (37.9%)	66 (37.9%)	29	13.24	0.01
Riesling	49 (28.2%)	72 (41.4%)	53 (30.5%)	29	10.41	0.01
Tintilla de Rota	41 (23.6%)	75 (43.1%)	58 (33.3%)	29	19.93	0.01

(% score in relation to the total).

Table 8. Triangular test alpha risk (α) results obtained in the comparative study with pollen and commercial activator.

Variety	Control vs. Pollen	Control vs. Commercial	Pollen vs. Commercial	Testers
Palomino Fino	n.d. (32.1%)	0.2 (42.9%)	0.01 (57.1%)	28
Riesling	0.2 (44.4%)	0.01 (66.7%)	0.2 (44.4%)	27
Tintilla de Rota	0.1 (50.0%)	0.2 (46.2%)	0.2 (46.2%)	26

n.d. (non-detected) (% tasters capable of recognizing the different wines).

According to results reported in Tables 7 and 8, we can conclude that there were consistent differences between the classification of wines:

- In Palomino Fino wines, according to classification test, the pollen and control wines were significantly preferred against the commercial activator. On the basis of a triangular test, we found that commercial activator wines were significantly different from the control ($\alpha = 0.2$) and the pollen wines ($\alpha = 0.01$). No significant differences were found between the control and pollen.
- In Riesling wines, pollen wines were significantly preferred over all others, distantly followed by the commercial activator and lastly by the control wines. According to the triangular test, significant differences were obtained in all comparisons.
- In Tintilla de Rota wines, regarding the classification test, similar results were obtained to Riesling, with a significant preference for pollen wines, followed by commercial activator and, lastly, control wines. On the basis of the results obtained from the triangular test, as in Riesling, we found that all the wine samples showed differences among themselves.

The largest percentage of tasters were able to identify the differences between the white wines. In Palomino Fino wines, this was by comparing pollen against the commercial activator, and in Riesling, between control and commercial.

According to the triangular test results for Tintilla de Rota red wines, we found that there was similar behavior of all the wines in general. The control–pollen confrontation obtained a slightly higher percentage than the rest.

4. Conclusions

Bee pollen use at a dose of 0.25 g/L versus a commercial activator led to an improvement in the formation of volatile compounds, especially higher alcohols, esters, and terpenes, resulting in an OAV increase of fruit and floral odor series.

Analyzing the classification and triangular test results together, along with the detailed information provided by the sensory analysis, we concluded that:

- The high percentages of tasters able to identify the different wines during the triangular test sessions proved the existence of significative differences amongst wines elaborated with pollen, commercial activator, and control.
- Results of the classification test showed a significant preference in most cases for the wines made with pollen, both in whites and reds.

Finally, taking into account the descriptive sensory analysis results, we were able to determine that these organoleptic differences between the wines were produced by improvements in the aromatic intensity of the sensory attributes corresponding to fruity and floral aromas. For this reason, the wines vinified with bee pollen were the best rated by the tasters.

Therefore, bee pollen could be considered as a valid and natural alternative to enhancing aromatic profile of fruits and flowers in young white and red wines.

Author Contributions: Conceptualization, A.A.-A., V.P. and A.J.-C.; data curation, A.A.-A. and V.P.; formal analysis, A.A.-A.; funding acquisition, A.A.-A. and V.P.; investigation, A.A.-A., P.S.-G., V.P. and A.J.-C.; methodology, A.A.-A., P.S.-G., V.P. and A.J.-C.; project administration, A.A.-A. and V.P.; supervision, V.P.; writing—original draft, A.A.-A., A.J.-C. and V.P.; writing—review and editing, A.A.-A., P.S.-G., A.J.-C. and V.P. All authors have read and agreed to the published version of the manuscript.

Funding: This research was funded by the University of Cadiz for a Ph.D. student assistantship under the program 2013-023/PU/EPIF-FPI-CT/CP.

Institutional Review Board Statement: Not applicable.

Informed Consent Statement: Not applicable.

Data Availability Statement: Not applicable.

Acknowledgments: A.A.-A. gratefully thanks the University of Cadiz for a Ph.D. student assistantship under the program 2013-023/PU/EPIF-FPI-CT/CP. The authors also thanks to the private wineries of Cádiz for supplying grapes.

Conflicts of Interest: The authors declare no conflict of interest.

References

1. Polášková, P.; Herszage, J.; Ebeler, S.E. Wine flavor: Chemistry in a glass. *Chem. Soc. Rev.* **2008**, *37*, 2478–2489. [CrossRef] [PubMed]
2. Niimi, J.; Danner, L.; Li, L.; Bossan, H.; Bastian, S.E.P. Wine consumers' subjective responses to wine mouthfeel and understanding of wine body. *Food Res. Int.* **2017**, *99*, 115–122. [CrossRef]
3. Vilanova, M.; Zamuz, S.; Fernando, V.; Sieiro, C. Effect of terroir on the volatiles of *Vitis vinifera* cv. Albariño. *J. Sci. Food Agric.* **2007**, *87*, 1252–1256. [CrossRef]
4. King, E.S.; Kievit, R.L.; Curtin, C.; Swiegers, J.H.; Pretorius, I.S.; Bastian, S.E.P.; Francis, I.L. The effect of multiple yeasts co-inoculations on Sauvignon Blanc wine aroma composition, sensory properties and consumer preference. *Food Chem.* **2010**, *122*, 618–626. [CrossRef]
5. Belda, I.; Ruiz, J.; Esteban-Fernández, A.; Navascués, E.; Marquina, D.; Santos, A.; Moreno-Arribas, M.V. Microbial contribution to Wine aroma and its intended use for Wine quality improvement. *Molecules* **2017**, *22*, 189. [CrossRef] [PubMed]
6. Álvarez-Pérez, J.M.; Campo, E.; San-Juan, F.; Coque, J.J.R.; Ferreira, V.; Hernández-Orte, P. Sensory and chemical characterisation of the aroma of Prieto Picudo rosé wines: The differential role of autochthonous yeast strains on aroma profiles. *Food Chem.* **2012**, *133*, 284–292. [CrossRef]
7. Rollero, S.; Bloem, A.; Ortiz-Julien, A.; Camarasa, C.; Divol, B. Fermentation performances and aroma production of non-conventional wine yeasts are influenced by nitrogen preferences. *FEMS Yeast Res.* **2018**, *18*, 1–11. [CrossRef]
8. Rollero, S.; Bloem, A.; Ortiz-Julien, A.; Camarasa, C.; Divol, B. Altered fermentation performances, growth, and metabolic footprints reveal competition for nutrients between yeast species inoculated in synthetic grape juice-like medium. *Front. Microbiol.* **2018**, *9*, 1–12. [CrossRef]
9. Bataillon, M.; Rico, A.; Sablayrolles, J.M.; Salmon, J.M.; Barre, P. Early thiamin assimilation by yeasts under enological conditions: Impact on alcoholic fermentation kinetics. *J. Ferment. Bioeng.* **1996**, *82*, 145–150. [CrossRef]
10. Bisson, L.F. Stuck and sluggish fermentations. *Am. J. Enol. Vitic.* **1999**, *50*, 107–119.
11. Blateyron, L.; Sablayrolles, J.M. Stuck and slow fermentations in enology: Statistical study of causes and effectiveness of combined additions of oxygen and diammonium phosphate. *J. Biosci. Bioeng.* **2001**, *91*, 184–189. [CrossRef]
12. Wang, X.D.; Bohlscheid, J.C.; Edwards, C.G. Fermentative activity and production of volatile compounds by *Saccharomyces* grown in synthetic grape juice media deficient in assimilable nitrogen and/or pantothenic acid. *J. Appl. Microbiol.* **2003**, *94*, 349–359. [CrossRef] [PubMed]
13. Bohlscheid, J.C.; Fellman, J.K.; Wang, X.D.; Ansen, D.; Edwards, C.G. The influence of nitrogen and biotin interactions on the performance of *Saccharomyces* in alcoholic fermentations. *J. Appl. Microbiol.* **2007**, *102*, 390–400. [CrossRef]
14. Walker, G.; Stewart, G. *Saccharomyces cerevisiae* in the Production of Fermented Beverages. *Beverages* **2016**, *2*, 30. [CrossRef]
15. Amores-Arrocha, A. *Aplicación de Polen de Abeja Como Activador en el Proceso de Fermentación Alcohólica*; Universidad de Cádiz: Cádiz, Spain, 2018.
16. Arias-Gil, M.; Garde-Cerdán, T.; Ancín-Azpilicueta, C. Influence of addition of ammonium and different amino acid concentrations on nitrogen metabolism in spontaneous must fermentation. *Food Chem.* **2007**, *103*, 1312–1318. [CrossRef]
17. Garde-Cerdán, T.; Ancín-Azpilicueta, C. Effect of the addition of different quantities of amino acids to nitrogen-deficient must on the formation of esters, alcohols, and acids during wine alcoholic fermentation. *LWT* **2008**, *41*, 501–510. [CrossRef]
18. Genovese, A.; Gambuti, A.; Piombino, P.; Moio, L. Sensory properties and aroma compounds of sweet Fiano wine. *Food Chem.* **2007**, *103*, 1228–1236. [CrossRef]
19. Hernández-Orte, P.; Ibarz, M.J.; Cacho, J.; Ferreira, V. Effect of the addition of ammonium and amino acids to musts of Airen variety on aromatic composition and sensory properties of the obtained wine. *Food Chem.* **2005**, *89*, 163–174. [CrossRef]
20. Kotarska, K.; Czupryński, B.; Kłosowski, G. Effect of various activators on the course of alcoholic fermentation. *J. Food Eng.* **2006**, *77*, 965–971. [CrossRef]
21. Bell, S.-J.; Henschke, P.A. Implications of nitrogen nutrition for grapes, fermentation and wine. *Aust. J. Grape Wine Res.* **2005**, *11*, 242–295. [CrossRef]
22. Lee, P.R.; Toh, M.; Yu, B.; Curran, P.; Liu, S.Q. Manipulation of volatile compound transformation in durian wine by nitrogen supplementation. *Int. J. Food Sci. Technol.* **2013**, *48*, 650–662. [CrossRef]
23. Trinh, T.; Woon, W.Y.; Yu, B.; Curran, P.; Liu, S.Q. Effect of L-isoleucine and L-phenylalanine Addition on Aroma Compound Formation During Longan Juice Fermentation by a Co-culture of *Saccharomyces cerevisiae* and *Williopsis saturnus*. *S. Afr. J. Enol. Vitic.* **2010**, *31*, 116–124. [CrossRef]
24. Barth, O.M.; Almeida-Muradian, L.B.; Pamplona, L.C.; Coimbra, S.; Barth, O.M. Chemical composition and botanical evaluation of dried bee pollen pellets. *J. Food Compos. Anal.* **2005**, *18*, 105–111.
25. Human, H.; Nicolson, S.W. Nutritional content of fresh, bee-collected and stored pollen of Aloe greatheadii var. davyana (Asphodelaceae). *Phytochemistry* **2006**, *67*, 1486–1492. [CrossRef]

26. Nicolson, S.W.; Human, H. Chemical composition of the "low quality" pollen of sunflower (*Helianthus annuus*, Asteraceae). *Apidologie* **2013**, *44*, 144–152. [CrossRef]
27. Paramás, A.M.G.; Bárez, J.A.G.; Marcos, C.C.; García-Villanova, R.J.; Sánchez, J.S. HPLC-fluorimetric method for analysis of amino acids in products of the hive (honey and bee-pollen). *Food Chem.* **2006**, *95*, 148–156. [CrossRef]
28. Xu, X.; Sun, L.; Dong, J.; Zhang, H. Breaking the cells of rape bee pollen and consecutive extraction of functional oil with supercritical carbon dioxide. *Innov. Food Sci. Emerg. Technol.* **2009**, *10*, 42–46. [CrossRef]
29. Dong, J.; Yang, Y.; Wang, X.; Zhang, H. Fatty acid profiles of 20 species of monofloral bee pollen from China. *J. Apic. Res.* **2017**, *54*, 503–511. [CrossRef]
30. De-Melo, A.A.M.; Estevinho, L.M.; Moreira, M.M.; Delerue-Matos, C.; Freitas, A.D.S.D.; Barth, O.M.; Almeida-Muradian, L.B.D. Phenolic profile by HPLC-MS, biological potential, and nutritional value of a promising food: Monofloral bee pollen. *J. Food Biochem.* **2018**, *42*, e12536. [CrossRef]
31. Amores-Arrocha, A.; Roldán, A.; Jiménez-Cantizano, A.; Caro, I.; Palacios, V. Evaluation of the use of multiflora bee pollen on the volatile compounds and sensorial profile of Palomino Fino and Riesling white young wines. *Food Res. Int.* **2018**, *105*, 197–209. [CrossRef] [PubMed]
32. Amores-Arrocha, A.; Sancho-Galán, P.; Jiménez-Cantizano, A.; Palacios, V. Bee Pollen Role in Red Winemaking: Volatile Compounds and Sensory Characteristics of Tintilla de Rota Warm Climate Red Wines. *Foods* **2020**, *9*, 981. [CrossRef] [PubMed]
33. Amores-Arrocha, A.; Roldán, A.; Jiménez-Cantizano, A.; Caro, I.; Palacios, V. Effect on White Grape Must of Multiflora Bee Pollen Addition during the Alcoholic Fermentation Process. *Molecules* **2018**, *23*, 1321. [CrossRef] [PubMed]
34. Sancho-Galán, P.; Amores-Arrocha, A.; Jiménez-Cantizano, A.; Palacios, V. Use of Multiflora Bee Pollen as a Flor Velum Yeast Growth Activator in Biological Aging Wines. *Molecules* **2019**, *24*, 1763. [CrossRef] [PubMed]
35. Amores-Arrocha, A.; Sancho-Galán, P.; Jiménez-Cantizano, A.; Palacios, V. Bee Pollen as Oenological Tool to Carry out Red Winemaking in Warm Climate Conditions. *Agronomy* **2020**, *10*, 634. [CrossRef]
36. ISO NORM. In *Sensory Analysis: Apparatus Wine Tasting Glass*; ISO: Geneva, Switzerland, 1977.
37. Jackson, R.S. *Wine Tasting: A Professional Handbook*; Academic Press: London, UK, 2009; p. 350.
38. ISO NORM. In *Sensory Analysis: Methodology. Ranking*; ISO: Geneva, Switzerland, 2006.
39. ISO NORM. In *Sensory Analysis: Methodology. Triangle Test*; ISO: Geneva, Switzerland, 2004.
40. Anzaldúa-Morales, A. *La Evaluación Sensorial de los Alimentos en Teoría y la Práctica*, 1st ed.; Acribia: Zaragoza, Spain, 1994; pp. 13–14.
41. Domínguez, C.; Guillén, D.A.; Barroso, C.G. Análisis de fenoles volátiles en vino fino de jerez, mediante GC-FID, utilizando SPE como preparación de muestras. In Proceedings of the VI Jornadas Científicas 2001 Grupos de Investigación Enológica, Valencia, Spain, 5–7 June 2001; p. 29.

Article

Comprehensive 2D Gas Chromatography with TOF-MS Detection Confirms the Matchless Discriminatory Power of Monoterpenes and Provides In-Depth Volatile Profile Information for Highly Efficient White Wine Varietal Differentiation

Igor Lukić [1,2,*], Silvia Carlin [3] and Urska Vrhovsek [3]

[1] Institute of Agriculture and Tourism, K. Huguesa 8, 52440 Poreč, Croatia
[2] Centre of Excellence for Biodiversity and Molecular Plant Breeding, Svetošimunska 25, 10000 Zagreb, Croatia
[3] Department of Food Quality and Nutrition, Research and Innovation Centre, Fondazione Edmund Mach (FEM), Via E. Mach 1, 38098 San Michele all'Adige, TN, Italy; silvia.carlin@fmach.it (S.C.); urska.vrhovsek@fmach.it (U.V.)
* Correspondence: igor@iptpo.hr; Tel.: +385-52-408-327

Received: 26 October 2020; Accepted: 28 November 2020; Published: 2 December 2020

Abstract: To differentiate white wines from Croatian indigenous varieties, volatile aroma compounds were isolated by headspace solid-phase microextraction (HS-SPME) and analyzed by comprehensive two-dimensional gas chromatography with time-of-flight mass spectrometry (GC×GC-TOF-MS) and conventional one-dimensional GC-MS. The data obtained were subjected to uni- and multivariate statistical analysis. The extra separation ability of the GC×GC second dimension provided additional in-depth volatile profile information, with more than 1000 compounds detected, while 350 were identified or tentatively identified in total by both techniques, which allowed highly efficient differentiation. A hundred and sixty one compounds in total were significantly different across monovarietal wines. Monoterpenic compounds, especially α-terpineol, followed by limonene and linalool, emerged as the most powerful differentiators, although particular compounds from other chemical classes were also shown to have notable discriminating ability. In general, Škrlet wine was the most abundant in monoterpenes, Malvazija istarska was dominant in terms of fermentation esters concentration, Pošip contained the highest levels of particular C_{13}-norisoprenoids, benzenoids, acetates, and sulfur containing compounds, Kraljevina was characterized by the highest concentration of a tentatively identified terpene γ-dehydro-ar-himachalene, while Maraština wine did not have specific unambiguous markers. The presented approach could be practically applied to improve defining, understanding, managing, and marketing varietal typicity of monovarietal wines.

Keywords: two-dimensional gas chromatography; one-dimensional; wine; volatile aroma compounds; multivariate analysis; cultivar; Croatia

1. Introduction

Aroma is among the most important attributes that drive the perception of wine sensory quality and varietal typicity by consumers. It results from the occurrence of many diverse odoriferous volatile compounds of different origin. Primary or varietal aroma compounds originate from grapes, secondary or fermentation aroma compounds are produced in fermentation, while tertiary aromas are formed during maturation [1–3]. The three groups mentioned are not so clearly divided: most of the precursors of volatile aroma compounds originate from grapes and are in one way or another affected by fermentation and/or aging [4]. The final wine aroma profile is a result of complex interactive effects between many sources of variability, such as variety [5], geographical position characterized by specific agroecological conditions [6,7], viticultural practices [8], harvest date [9], harvest year [10,11], grape processing, and fermentation parameters [12,13], etc.

Varietal characterization (description) and differentiation (contradistinction from other varieties) is an ever-important field of wine research. Many studies have aimed to identify volatile compounds characteristic for various grape varieties, since they are crucial for the typical varietal attributes of their wines. The knowledge on the volatile aroma compound composition of monovarietal wines is important since it may enable producers to better cope with the phenomena encountered in production and to manage vinification with greater efficiency, all in order to produce high quality wines of accentuated varietal typicity. It may enable detailed and precise description of the aroma of monovarietal wines, which could be used in their marketing, especially towards informed consumers interested in wines of high quality with marked diversity and identity. In addition to often being linked to a given geographical provenance with a corresponding protected designation of origin (PDO), particular monovarietal wines are especially appreciated and demanded because of their typical sensory properties. Such wines often fall within a higher price range and are a target of counterfeiting by mislabeling their varietal origin. Therefore, control in terms of varietal origin authentication is needed: the general strategy used by many research groups includes the (semi)quantification of a large number of volatile compounds in large sets of wines and use of the generated data for the production of multivariate statistical models able to classify wines, as well as to predict and confirm their varietal origin [5].

The analysis of volatile aroma compounds in wine varietal characterization and differentiation studies is commonly performed by conventional one-dimensional gas chromatography mass spectrometry (GC-MS) [14–19]. Although the information obtained by this approach is often sufficient to obtain more or less efficient varietal differentiation, a large amount of information is lost due to frequent co-elutions, even when using long GC run times on high-efficiency capillary columns with selective stationary phases and programmed oven temperature conditions [20,21]. In the last few decades, comprehensive two-dimensional gas chromatography-mass spectrometry (2D-GC-MS or GC×GC-MS) stood out as a highly potent technique for in-depth characterization of complex samples [22], where the number of compounds of interest is large and many are present at trace levels, as in wine. This technique utilizes two GC columns of different stationary phases serially connected by a modulator, where the compounds co-eluting in the first column are in most cases separated in the second. GC×GC-MS is therefore characterized by higher efficiency and sensitivity, since the additional separation by a second stationary phase produces clearer mass spectra and much less chromatographic peaks remain unannotated. In this way, GC×GC-MS allows detection and identification of a much larger number of volatile compounds compared to conventional GC-MS [23].

Regardless of the existing great potential, only a few studies have utilized GC×GC to investigate wine volatile aroma profiles, while studies which used GC×GC for varietal characterization and differentiation were extremely rare. Several authors reported more or less detailed GC×GC volatile aroma profiles of particular monovarietal wines, such as Cabernet Sauvignon [24], Sauvignon Blanc [25], Shiraz [9,26] or Syrah [12], Pinotage [21], Chardonnay [27], and Verdicchio [28], but none of them directly compared them to or differentiated them from other monovarietal wines of similar typology. In this way, despite detailed profiles determined in some cases, it still remained unknown which compounds and in which amounts are typical for a given variety and whether they could differentiate it from other monovarietal wines. The only two studies which utilized GC×GC and succeeded in differentiating several monovarietal wines did not report actual concentrations of all the identified volatile compounds [20,29].

The aim of this study was to utilize the potential of two-dimensional gas chromatography with time-of-flight mass spectrometry (GC×GC-TOF-MS) technique, in combination with headspace solid-phase microextraction (HS-SPME) and multivariate statistical tools, as a more efficient approach to characterize and differentiate monovarietal white wines based on their volatile aroma compound composition. Profiling by GC×GC was combined with conventional GC-MS analysis of major wine volatile compounds to obtain more comprehensive aroma profiles. Special attention was devoted to terpenes, often highlighted as key varietal markers in wine. The approach was applied to characterize and differentiate Croatian wines made from indigenous grape varieties, with each variety represented by a rather heterogeneous group of wines with respect to geographical microlocation and agroecological conditions, viticultural practices, harvest date, and grape processing and wine production parameters. It was expected that GC×GC-TOF-MS would be extremely effective in providing novel in-depth information for efficient white wine varietal differentiation.

2. Materials and Methods

2.1. Wine Samples

A total of 32 wines made from Croatian indigenous white grape varieties (*Vitis vinifera* L.) Malvazija istarska (MI, 8 samples) Pošip (PO, 7), Maraština (MA, 7), Kraljevina (KR, 7), and Škrlet (SK, 3) were donated by producers from Croatia (EU), more specifically Istria (MI) and Dalmatia (PO and MA) as the coastal regions and continental Croatia (KR and SK). Wines from the same variety were donated by different producers. The selection was representative for Croatian wine production and comprised the majority of the most important Croatian indigenous varieties. Only young wines from harvest 2015 were collected, labelled with a protected designation of origin (PDO) and with a traditional term "Quality or Top quality" wine. Wines were of the same typology and produced by standard white winemaking technology, which included grape harvest at technological maturity, destemming, crushing and mashing of the grapes, no or short pre-fermentative skin-contact (up to 48 h), use of selected commercial yeasts, fermentation at relatively low temperatures (up to 18 °C), and other standard procedures (sulfiting, racking, fining, and stabilization, etc.). Wines were not in contact with wood. During the period from harvest and vinification in September 2015 until the collection and analyses in April and May 2016 the wines were stored in stainless steel tanks and 0.75 L glass bottles with cork stoppers in wine cellars of the producers. The wine samples were selected from a larger set as typical representatives of a given variety by the panel for wine sensory analysis of the Institute of Agriculture and Tourism in Poreč (Croatia), which consisted of highly trained and experienced tasters. Standard physico-chemical parameters of the collected wines determined by OIV methods are reported in Table S1.

2.2. Standards, Chemicals, and Consumables

Chemical standards of volatile aroma compounds were procured from AccuStandard Inc. (New Haven, CT, USA), Fluka (Buchs, Switzerland), Honeywell International Inc. (Morris Plains, NJ, USA), Merck (Darmstadt, Germany), and Sigma-Aldrich (Sigma-Aldrich, St. Louis, MO, USA). A stock solution of major volatile compounds commonly present in wine was prepared in methanol, while standard solutions were prepared in model wine (13 vol.% of ethanol, pH 3.3). Ammonium sulfate and sodium chloride were purchased from Kemika d.d (Zagreb, Croatia).

Divinylbenzene/carboxen/polydimethylsiloxane (DVB-CAR-PDMS, StableFlex, 50/30 μm, 1 cm) SPME fiber used for GC-MS analysis was procured from Supelco, Sigma Aldrich (Bellafonte, PA, USA) and DVB-CAR-PDMS SPME fiber (StableFlex, 50/30 μm, 2 cm) used for GC×GC-TOF-MS analysis was procured from Supelco, Sigma Aldrich (Milan, Italy).

2.3. Analysis of Volatile Aroma Compounds by Conventional One-Dimensional GC-MS

Volatile aroma compounds for GC-MS analysis were isolated by headspace solid-phase microextraction (HS-SPME) according to the modified method proposed by Bubola et al. [30]. Four milliliters of a solution obtained by diluting wine four times with deionized water were pipetted in a 10 mL glass vial. Ammonium sulfate (1 g) and 50 μL of internal standards solution (2-octanol (0.84 mg/L), 1-nonanol (0.82 mg/L), and heptanoic acid (2.57 mg/L)) were added. After 15 min preconditioning at 40 °C, microextraction using a DVB-CAR-PDMS SPME fiber took place for 40 min at 40 °C with stirring (800 rpm). Volatile compounds were desorbed after the insertion of the fiber for 10 min into a GC/MS injector heated at 248 °C, with the first 3 min in splitless mode. Volatile aroma compounds were identified and quantified using a Varian 3900 gas chromatograph (GC) connected to a Varian Saturn 2100T mass spectrometer with an ion trap analyzer (Varian Inc., Harbour City, CA, USA). The column used was a 60 m × 0.25 mm i.d. × 0.25 μm d.f. Rtx-WAX (Restek, Belafonte, PA, USA). Initial temperature of the GC oven was 40 °C, ramped up at 2 °C/min to reach 240 °C, and then kept at this temperature for additional 10 min. Helium was used as a carrier gas at a flow rate of 1.2 mL/min. Mass spectra were acquired in EI mode (70 eV), at 30–350 *m/z*.

Identification of volatile compounds was conducted by comparison of retention times and mass spectra of the analytes with those of pure standards, and with mass spectra from NIST05 library. Identification by comparison with mass spectra was considered satisfactory if spectra reverse match numbers (RM) higher than 800 were obtained. In the case of less clear spectra (RM < 800) identification was considered satisfactory if the ratios of the relative intensities of a quantifier ion and three characteristic ions with the highest intensity reasonably matched those in the reference spectra of a given compound. Linear retention indices were calculated with respect to the retention times of C_{10} to C_{28} n-alkanes and compared to those reported in literature for columns of equal or equivalent polarity. Calibration curves were constructed based on the analysis of standard solutions containing known concentrations of standards at six concentration levels and were used for quantification. Quantification of major volatile compounds was based on total ion current peak area, while quantification of minor compounds was based on quantifier ion peak area. The peak areas and concentrations in standard solutions and in wine samples were normalized with respect to those of the internal standards. Linearity was satisfactory with coefficient of determination higher than 0.99 for all the standards. Relative standard deviation of repeatability (RSD) was determined after repeated analysis ($n = 5$) of a Malvazija istarska wine sample and was satisfactory, with RSD lower than 13.05% for monoterpenes, 7.38 for β-damasenone, lower than 9.23% for alcohols, 7.34 for ethyl esters, 12.34% for acetate esters, and 11.78% for fatty acids. Method validation parameters were previously published in the study of Bubola et al. [30]. In the cases when pure chemical standards were not available, semi-quantitative analysis was carried out. The concentrations of such compounds were expressed as equivalents of compounds with similar chemical structure which were quantified using calibration curves, assuming a response factor equal to one.

2.4. Analysis of Volatile Aroma Compounds by GC×GC-TOF-MS

A volume of 2.5 mL of wine was transferred to a 20 mL headspace vial and 1.5 g of sodium chloride was added. Wine sample was spiked with 50 μL of internal standard (2-octanol, 1 mg/L). Quality control samples (QC) were prepared by mixing equal proportion of each sample and were analyzed before the samples sequence ($n = 5$) and after every five samples ($n = 1$). GC×GC-TOF-MS analysis of wines was performed using a GC Agilent 7890N (Agilent Technologies, Palo Alto, CA, USA) coupled to a LECO Pegasus IV time-of-flight mass spectrometer (TOF-MS) (Leco Corporation, St. Joseph, MI, USA) equipped with a Gerstel MPS autosampler (GERSTEL GmbH & Co. KG, Mülheim an der Ruhr, Germany), as described in previous studies with minor modifications [9,31,32]. Briefly, samples were preconditioned at 35 °C for 5 min and volatile compounds were extracted using a DVB/CAR/PDMS SPME fiber for 20 min. Volatile compounds were desorbed for 3 min at 250 °C in splitless mode. The fiber was reconditioned for 7 min at 270 °C between each extraction. Helium was used as a carrier gas at a flow rate of 1.2 mL/min. The oven was equipped with a 30 m × 0.25 mm × 0.25 μm film thickness VF-WAXms column (Agilent Technologies) in the first dimension (1D) and a 1.5 m × 0.15 mm × 0.15 μm film thickness Rxi 17Sil MS column (Restek) in the second dimension (2D). Initial oven temperature was maintained at 40 °C for 4 min, then raised at 6 °C/min to 250 °C, and then finally maintained at this temperature for additional 5 min. The second oven was maintained at 5 °C above the temperature of the first one throughout the analysis. The modulator was offset by +15 °C in relation to the secondary oven, the modulation time was 7 s with 1.4 s of hot pulse duration, as described previously [31]. Electron ionization at 70 eV was applied, the temperature of ion source was 230 °C, detector voltage was 1317 V, mass range (*m/z*) was 40–350, acquisition rate was 200 spectra/s, and acquisition delay was 120 s.

Baseline correction, chromatogram deconvolution and peak alignment were performed using LECO ChromaTOF software version 4.32 (Leco Corporation, St. Joseph, MI, USA). The baseline offset was set to 0.8 and signal to noise (S/N) ratio was set at 100. Peak width limits were set to 42 s and 0.1 s in the first and the second dimension, respectively. Traditional, not adaptive integration was used. The required match (similarity) to combine peaks was set to 650. Under these conditions 1025 putative compounds were detected. Volatile compounds were identified by comparing their retention times and mass spectra with those of pure standards and with mass spectra from NIST 2.0, Wiley 8, and FFNSC 2 (Chromaleont, Messina, Italy) mass spectral libraries, with a minimum library similarity match factor of 750 out of 999. For identification of compounds by comparison with pure standards, a mix of 122 compounds was injected under identical GC×GC-TOF-MS conditions. For tentative identification of compounds and/or confirmation of their identities determined as described above, linear retention indices were calculated with respect to the retention times of C_{10} to C_{30} n-alkanes and compared to those from literature for conventional one-dimensional GC obtained using columns of equal or equivalent polarity (NIST 2.0, Wiley 8, FFNSC 2, VCF, ChemSpider). Three hundred and seventeen (317) volatile aroma compounds were (tentatively) identified in total. Volatile compounds were semi-quantified and their concentrations in μg/L were calculated relative to the internal standard 2-octanol, assuming a response factor equal to one.

In preliminary tests by principal component analysis (PCA), QC samples were clustered very close and were very well separated from the wine samples, suggesting the repeatability of the method was very good. Relative standard deviation of the internal standard 2-octanol in QC samples was 10.4% which was considered satisfactory for HS-SPME/GC×GC-TOF-MS analysis.

2.5. Statistical Data Elaboration

Data obtained by GC-MS and GC×GC-TOF-MS were processed by analysis of variance (one-way ANOVA). Least significant difference (LSD) post-hoc test was used to compare the mean values of concentrations at $p < 0.05$. Multivariate analysis of data was performed by PCA and forward stepwise linear discriminant analysis (SLDA). The original dataset which included 32 wines and 350 volatile aroma compounds (33 determined by GC-MS + 317 determined by GC×GC-TOF-MS analysis; in the case of compounds determined by both techniques GC×GC-TOF-MS data were used), was reduced based on Fisher ratios (*F*-ratios). Multivariate techniques were applied on the variables (mean-centered concentrations of volatile compounds) with the highest *F*-ratios. PCA was performed with 40 variables with the highest *F*-ratio, while SLDA and hierarchical clustering were performed with 60 variables with the highest *F*-ratio, in both cases with GC-MS and GC×GC-TOF-MS data combined. Two additional SLDA models were built with the concentrations of terpenes which were significantly different between wines, using GC-MS and GC×GC-TOF-MS data separately. In SLDA, variables were selected based on Wilk's lambda, with *F* to enter = 1 and *F* to remove = 0.5. Cross-validation was applied to check the prediction capacity of the developed SLDA models. ANOVA, PCA, and SLDA were performed by Statistica v. 13.2 software (StatSoft Inc., Tulsa, OK, USA). Hierarchical clustering was conducted and a heatmap was generated by Ward algorithm and Euclidean distance analysis using MetaboAnalyst v. 4.0 (http://www.metaboanalyst.ca), created at the University of Alberta, Canada [33].

3. Results and Discussion

3.1. GC-MS

Major volatile aroma compounds are highly abundant in wines and for this reason GC-MS was considered appropriate for their analysis. It was considered that their quantitation by GC-MS was not significantly affected by co-eluting compounds. As well, the analysis of major volatiles by GC×GC-TOF-MS would require a rather different setup than that applied in this study, with much larger modulation time and hot pulse duration, not applicable for minor and trace compounds. Major volatile aroma compounds determined by GC-MS are listed in Table 1, grouped according to chemical class, and sorted within each class in order of decreasing *F*-ratio obtained by one-way ANOVA. Twenty-one monoterpenoids and a sesquiterpenoid *trans*-nerolidol, eight C_{13}-norisoprenoids, two benzenoids, four alcohols, four acids, and 11 esters were quantified. Table S2 reports the concentrations of the identified volatile compounds in each of the investigated wines.

Among terpenes, major monoterpenols such as linalool, geraniol, α-terpineol, and nerol were found in the highest concentration, which was generally in agreement with previous findings on white wines [34–36]. The mentioned are among the most influential monoterpenoids to wine aroma, to which they significantly contribute with specific floral and fruity nuances due to their relatively low odor perception thresholds, such as, for example, 15 μg/L for linalool [35,37]. The highest *F*-ratio among all the compounds identified by GC-MS was determined for α-terpineol, followed by an unidentified monoterpene and linalool, confirming the importance of terpenes for wine varietal differentiation [35]. Many other (mono)terpenes also turned out to be important in this sense, while other compound classes exhibited lower *F*-ratios, with the exception of 1-hexanol. Such an outcome was expected to some extent, since terpenes are primary aroma compounds originating from grapes, both as free volatile molecules or released from glycosidic precursors. Their composition and amounts are genetically pre-determined: genetic variation in aroma biosynthesis genes cause differences in terpene concentrations between grapevine varieties. For example, a variant of 1-deoxy-D-xylulose-5-phosphate synthase, a gene responsible for the biosynthesis of terpenoids, causes pronounced increase in terpene concentration in Muscat and Gewürztraminer grapes, which gives wines of these varieties a recognizable floral aroma [4,38,39]. Monoterpenes are generally known to be responsible for varietal aroma of muscats and non-muscat aromatic varieties, such as Gewürtztraminer, Riesling, Müller-Thurgau, etc. [36,40,41], but were also found useful for the differentiation of wines of other, so-called semi-aromatic and neutral grape varieties [41–45]. Márquez, Castro, Natera, and García-Barroso [46] characterized the volatile fraction of Andalusian sweet wines made from Muscat and Pedro Ximenez varieties and, interestingly, also found that α-terpineol was the most powerful differentiator with the highest *F*-ratio, followed closely by linalool and limonene, similar as in this case.

In this study, the ratios of terpene concentrations in different monovarietal wines varied from compound to compound, but it was generally observed that wines from Škrlet, a relatively unexplored Croatian grape variety, were characterized by the highest concentrations of many important monoterpenes (Table 1), while the concentrations of other monoterpenes were also among the highest in the investigated wines. The concentrations of monoterpenes in Malvazija istarska wines were notable and generally in fair agreement with those reported previously for this variety, with linalool followed by geraniol as the most abundant [43,47–49]. Malvazija was followed by Pošip wine with intermediate concentrations, while Maraština and especially Kraljevina wines had the lowest terpene concentrations.

Table 1. Concentrations (μg/L) of volatile aroma compounds found in Croatian monovarietal wines after headspace solid-phase microextraction followed by gas chromatography-mass spectrometry (HS-SPME/GC–MS) sorted by compound class and descending Fisher *F*-ratio.

No.	Volatile Compounds	t_R	ID	LRI_{exp}	LRI_{lit}	*F*-Ratio	Variety				
		(min:s)					MI	PO	MA	KR	SK
	Terpenes										
1	α-Terpineol	39:59	S, MS, LRI	1684	1684	35.07	15.65 ± 7.04 ^{b}	10.92 ± 3.16 ^{bc}	5.50 ± 2.10 ^{cd}	1.98 ± 0.89 ^{d}	40.49 ± 11.05 ^{a}
2	Monoterpene (n.i.; *m/z* 59, 93, 121)	28:29	MS	1441	-	27.68	1.03 ± 0.50 ^{b}	0.64 ± 0.24 ^{bc}	0.27 ± 0.24 ^{cd}	0.02 ± 0.04 ^{d}	2.79 ± 1.01 ^{a}
3	Linalool	33:10	S, MS, LRI	1542	1542	24.71	68.00 ± 27.76 ^{b}	38.17 ± 8.05 ^{c}	18.52 ± 8.28 ^{d}	7.64 ± 2.56 ^{d}	90.75 ± 15.48 ^{a}
4	Limonene	15:17	MS, LRI	1191	1196	18.91	1.33 ± 0.68 ^{b}	0.98 ± 0.28 ^{b}	0.19 ± 0.05 ^{c}	0.36 ± 0.11 ^{c}	2.66 ± 1.01 ^{a}
5	Nerol	44:35	S, MS, LRI	1791	1791	16.31	13.38 ± 6.72 ^{a}	6.49 ± 1.96 ^{b}	4.55 ± 1.58 ^{bc}	1.06 ± 0.46 ^{c}	17.36 ± 3.74 ^{a}
6	*cis*-Linalool furan oxide	29:24	MS, LRI	1464	1464	12.57	0.08 ± 0.03 ^{b}	0.18 ± 0.06 ^{a}	0.06 ± 0.05 ^{b}	0.02 ± 0.01 ^{b}	0.20 ± 0.11 ^{a}
7	Monoterpenyl acetate (n.i.; *m/z* 93, 69, 121)	21:55	MS	1302	-	12.50	3.12 ± 1.85 ^{a}	1.11 ± 0.27 ^{b}	0.59 ± 0.36 ^{b}	0.12 ± 0.06 ^{b}	3.16 ± 0.84 ^{a}
8	4-Terpineol	35:37	MS, LRI	1594	1596	11.60	0.24 ± 0.10 ^{b}	0.24 ± 0.06 ^{b}	0.23 ± 0.11 ^{b}	0.11 ± 0.03 ^{c}	0.51 ± 0.08 ^{a}
9	β-Pinene	13:45	MS, LRI	1146	1145	11.46	4.40 ± 2.41 ^{a}	2.43 ± 0.75 ^{bc}	0.40 ± 0.15 ^{d}	1.16 ± 0.37 ^{cd}	4.17 ± 1.06 ^{ab}
10	Ho-Trienol	36:02	MS, LRI	1601	1601	10.44	7.45 ± 3.95 ^{a}	6.95 ± 2.01 ^{ab}	1.71 ± 1.14 ^{c}	1.60 ± 0.76 ^{c}	4.18 ± 0.79 ^{bc}
11	*trans*-Rose oxide	23:23	MS, LRI	1352	1341	10.32	0.27 ± 0.08 ^{b}	0.21 ± 0.04 ^{bc}	0.15 ± 0.07 ^{c}	0.16 ± 0.03 bc	0.59 ± 0.34 ^{a}
12	Monoterpene (n.i.; *m/z* 93, 69, 41)	29:56	MS	1476	-	8.51	0.49 ± 0.28 ^{b}	0.25 ± 0.07 ^{c}	0.24 ± 0.17 ^{c}	0.12 ± 0.13 c	0.77 ± 0.24 ^{a}
13	*trans*-Ocimene	18:03	MS, LRI	1252	1250	8.45	1.58 ± 0.92 ^{a}	1.27 ± 0.41 ^{ab}	0.17 ± 0.07 ^{c}	0.55 ± 0.18 ^{bc}	1.30 ± 0.50 ^{ab}
14	Citronellol	43:11	S, MS, LRI	1758	1758	8.34	5.02 ± 0.61 ^{a}	5.09 ± 0.69 ^{a}	5.30 ± 1.78 ^{a}	2.56 ± 0.30 ^{b}	5.60 ± 1.75 ^{a}
15	Nerol oxide	29:18	MS, LRI	1459	1464	7.01	3.04 ± 1.12 ^{a}	3.74 ± 1.82 ^{a}	1.35 ± 1.17 ^{b}	1.11 ± 0.40 ^{b}	4.11 ± 1.44 ^{a}
16	Geranyl acetone	47:01	MS, LRI	1845	1845	5.38	2.93 ± 0.58 ^{b}	3.58 ± 0.99 ^{b}	7.58 ± 4.80 ^{a}	2.64 ± 0.39 ^{b}	2.55 ± 1.14 ^{b}
17	*trans*-Linalool pyran oxide	41:49	MS, LRI	1726	1752	4.85	0.08 ± 0.02 ^{b}	0.13 ± 0.05 ^{a}	0.07 ± 0.05 ^{b}	0.04 ± 0.03 ^{b}	0.06 ± 0.02 ^{b}
18	*trans*-Nerolidol	54:39	MS, LRI	2031	2031	4.61	2.89 ± 0.50 ^{a}	3.17 ± 0.59 ^{a}	2.66 ± 1.58 ^{ab}	1.59 ± 0.22 ^{b}	1.53 ± 0.35 ^{b}
19	Monoterpene (n.i.; *m/z* 121, 93, 136)	31:30	MS	1509	-	3.09	2.45 ± 0.49 ^{a}	2.41 ± 0.56 ^{a}	2.11 ± 0.73 ^{a}	1.11 ± 0.16 ^{b}	2.88 ± 2.79 ^{a}
20	Geraniol	46:35	S, MS, LRI	1838	1838	2.93	40.64 ± 21.59 ^{ab}	24.23 ± 8.96 ^{ab}	39.96 ± 48.27 ^{ab}	2.73 ± 1.56 ^{b}	46.19 ± 10.53 ^{ab}
21	Geranyl ethyl ether	31:54	MS, LRI	1511	1499	2.69	0.53 ± 0.33	0.86 ± 0.97	1.08 ± 0.84	0.05 ± 0.02	0.82 ± 0.25
22	α-Terpinolene	19:34	MS, LRI	1287	1281	2.32	0.49 ± 0.29	0.73 ± 0.92	0.07 ± 0.04	0.14 ± 0.07	0.33 ± 0.26
	C_{13}-norisoprenoids										
23	Vitispirane II	31:16	MS, LRI	1523	1529	9.85	0.07 ± 0.02 ^{c}	0.34 ± 0.16 ^{a}	0.20 ± 0.10 ^{b}	0.09 ± 0.01 ^{c}	0.14 ± 0.06 ^{bc}
24	β-Damascenone	45:26	MS, LRI	1809	1809	7.09	3.52 ± 0.69 ^{a}	2.81 ± 1.42 ^{ab}	1.99 ± 0.58 ^{bc}	2.28 ± 0.25 ^{b}	0.89 ± 0.29 ^{c}
25	Actinidol I	49:55	MS, LRI	1914	1914	5.59	0.12 ± 0.05 ^{a}	0.16 ± 0.06 ^{a}	0.13 ± 0.07 ^{a}	0.04 ± 0.01 ^{b}	0.09 ± 0.03 ^{ab}
26	Actinidol II	50:27	MS, LRI	1927	1927	5.10	0.20 ± 0.08 ^{a}	0.23 ± 0.07 ^{a}	0.23 ± 0.10 ^{a}	0.08 ± 0.01 ^{b}	0.16 ± 0.04 ^{ab}
27	Vitispirane I	32:08	MS, LRI	1521	1526	5.03	0.09 ± 0.04 ^{c}	0.46 ± 0.24 ^{a}	0.33 ± 0.24 ^{ab}	0.19 ± 0.05 ^{bc}	0.32 ± 0.18 ^{abc}
28	β-Ionone	50:17	S, MS, LRI	1923	1923	3.89	0.06 ± 0.01 ^{ab}	0.05 ± 0.01 ^{b}	0.07 ± 0.01 ^{a}	0.05 ± 0.01 ^{b}	0.07 ± 0.01 ^{a}
29	Actinidol ethyl ether I	40:25	MS, LRI	1690	1690	3.37	0.25 ± 0.12 ^{bc}	0.43 ± 0.24 ^{a}	0.34 ± 0.25 ^{ab}	0.11 ± 0.02 ^{c}	0.24 ± 0.06 ^{bc}

Table 1. *Cont.*

No.	Volatile Compounds	t_R (min:s)	ID	LRI_{exp}	LRI_{lit}	*F*-Ratio	Variety				
							MI	PO	MA	KR	SK
30	Actinidol ethyl ether II	41:49	MS, LRI	1723	1723	2.76	0.15 ± 0.07 ^{ab}	0.25 ± 0.16 ^{a}	0.20 ± 0.15 ^{a}	0.06 ± 0.01 ^{b}	0.15 ± 0.04 ^{ab}
	Benzenoids										
31	Ethyl cinnamate	57:33	S, MS, LRI	2111	2122	6.96	0.41 ± 0.19 ^{b}	1.16 ± 0.78 ^{a}	0.39 ± 0.08 ^{b}	0.21 ± 0.10 ^{b}	0.16 ± 0.10 ^{b}
32	Benzaldehyde	31:26	S, MS, LRI	1508	1509	0.84	1.66 ± 1.25	3.48 ± 5.40	1.17 ± 0.54	2.56 ± 0.81	3.11 ± 1.57
	Alcohols										
33	1-Hexanol	23:35	S, MS, LRI	1356	1357	25.56	792.14 ± 264.44 ^{b}	949.93 ± 179.86 ^{b}	859.15 ± 171.18 ^{b}	321.89 ± 32.90 ^{c}	1544.09 ± 146.31 ^{a}
34	*cis*-3-Hexen-1-ol	25:03	S, MS, LRI	1379	1379	12.73	77.49 ± 40.64 ^{c}	299.33 ± 113.23 ^{a}	193.20 ± 123.23 ^{b}	26.16 ± 4.63 ^{c}	54.67 ± 23.77 ^{c}
35	2-Phenylethanol	48:52	S, MS, LRI	1891	1893	7.16	20,047.0 ± 4767.1 ^{b}	33,176.1 ± 4679.3 ^{a}	32,117.2 ± 10,870.7 ^{a}	20,712.5 ± 6134.8 ^{b}	17,665.9 ± 1061.0 ^{b}
36	*trans*-3-Hexen-1-ol	24:03	S, MS, LRI	1361	1361	1.73	61.38 ± 24.09	45.64 ± 17.99	46.57 ± 10.28	43.09 ± 10.26	63.87 ± 22.80
	Acids										
37	Decanoic acid	62:49	S, MS, LRI	2257	2258	5.05	646.02 ± 179.70 ^{b}	1627.60 ± 659.33 ^{a}	1062.71 ± 505.33 ^{b}	994.33 ± 67.19 ^{b}	1090.36 ± 494.95 ^{ab}
38	Octanoic acid	54:56	S, MS, LRI	2043	2042	4.03	4294.07 ± 796.78 ^{b}	6239.74 ± 1532.91 ^{a}	5147.23 ± 1562.12 ^{ab}	6219.42 ± 455.69 ^{a}	6359.73 ± 1152.33 ^{a}
39	Hexanoic acid	46:10	S, MS, LRI	1830	1828	3.05	5715.09 ± 552.13 ^{ab}	5184.65 ± 722.46 ^{b}	5284.54 ± 1710.50 ^{b}	6487.89 ± 603.01 ^{a}	7025.45 ± 1103.35 ^{a}
40	Butyric acid	36:28	S, MS, LRI	1612	1612	0.54	1766.10 ± 323.75	1607.09 ± 231.34	1685.41 ± 407.86	1581.53 ± 184.63	1788.32 ± 346.09
	Esters										
41	2-Phenethyl acetate	45:03	S, MS, LRI	1803	1801	9.02	2230.06 ± 481.79 ^{b}	4731.20 ± 1467.85 ^{a}	2359.08 ± 1289.62 ^{b}	2579.92 ± 287.25 ^{b}	1750.70 ± 284.91 ^{b}
42	Ethyl octanoate	28:06	S, MS, LRI	1435	1435	8.88	1211.04 ± 239.22 ^{a}	1086.51 ± 223.88 ^{a}	817.08 ± 231.10 ^{b}	701.64 ± 160.66 ^{b}	544.02 ± 243.59 ^{b}
43	Ethyl hexanoate	17:35	S, MS, LRI	1236	1236	6.80	721.60 ± 172.38 ^{a}	379.34 ± 86.89 ^{c}	463.42 ± 153.50 ^{bc}	580.60 ± 120.60 ^{ab}	474.95 ± 108.08 ^{bc}
44	Hexyl acetate	19:26	S, MS, LRI	1272	1272	6.10	216.64 ± 52.04 ^{a}	204.45 ± 73.60 ^{a}	123.25 ± 54.35 ^{b}	107.91 ± 34.65 ^{b}	207.09 ± 40.86 ^{a}
45	Ethyl decanoate	37:43	S, MS, LRI	1637	1638	5.61	302.58 ± 46.92 ^{a}	279.95 ± 69.24 ab	179.10 ± 81.94 ^{c}	220.08 ± 29.45 ^{bc}	199.26 ± 29.89 ^{bc}
46	Isoamyl acetate	12:29	S, MS, LRI	1120	1122	3.97	3299.12 ± 1092.74 ^{a}	3321.37 ± 1674.71 ^{a}	1460.92 ± 566.57 ^{b}	2397.45 ± 774.95 ^{ab}	1879.81 ± 562.20 ^{ab}
47	Ethyl butyrate	09:27	S, MS, LRI	1030	1030	3.09	456.83 ± 69.21 ^{a}	415.44 ± 50.58 ^{ab}	363.29 ± 80.70 ^{b}	367.30 ± 47.60 ^{b}	350.99 ± 74.33 ^{b}
48	Ethyl 3-methylbutyrate	10:31	S, MS, LRI	1065	1065	2.58	8.51 ± 1.98	14.79 ± 5.57	12.86 ± 5.26	11.34 ± 4.39	8.39 ± 0.88
49	Ethyl 2-methylbutyrate	10:00	S, MS, LRI	1049	1049	2.17	4.19 ± 1.13	6.57 ± 2.31	6.57 ± 2.70	6.25 ± 2.48	4.01 ± 0.62
50	Diethyl succinate	39:04	S, MS, LRI	1667	1669	1.53	1634.59 ± 398.33	1917.40 ± 1362.67	1665.03 ± 858.64	997.49 ± 290.90	1064.55 ± 106.44
51	Ethyl lactate	22:56	S, MS, LRI	1341	1341	0.58	25,943.4 ± 13,586.8	45,815.9 ± 55,981.7	34,462.1 ± 16,552.6	25,359.5 ± 12,701.8	32654.3 ± 7282.3

ID—identification of compounds; S—retention time and mass spectrum consistent with that of the pure standard and with NIST05 mass spectra electronic library; LRI—linear retention index consistent with that found in literature; MS—mass spectra consistent with that from NIST05 mass spectra electronic library or literature; n.i.—not identified. The compounds with only MS symbol in ID column were tentatively identified. The compounds for which pure standards were not available (without symbol S in the ID column) were quantified semi-quantitatively and their concentrations were expressed as equivalents of compounds with similar chemical structure assuming a response factor = 1. LRI_{exp}—linear retention index obtained experimentally. Varieties: MI—Malvazija istarska, PO—Pošip, MA—Maraština, KR—Kraljevina, SK—Škrlet. Different superscript lowercase letters in a row represent statistically significant differences between mean values at $p < 0.05$ obtained by one-way ANOVA and least significant difference (LSD) test.

Although the content and composition of terpenes in grapes and wines is principally pre-determined by variety, they are susceptible to modulation in response to many factors, such as viticultural parameters including soil characteristics, exposure to sunlight, water status, defoliation, crop thinning, etc. [34,50], as well as pre-fermentation and fermentation practices and conditions [35,36]. Except the effect of variety, the differences between the investigated monovarietal wines were probably partly caused by different geographical origin (Istria, Dalmatia, continental Croatia), so the effects of variety and location probably acted in synergy. It is indeed known that low temperatures favor the production of aroma compounds in grapes [51], so it is possible that the highest concentration of monoterpenes in Škrlet wines from continental Croatia characterized by lower temperatures was at least partly due to the effect of climate. The same could be deduced for Malvazija wines coming from the northern, somewhat colder part of the Adriatic coast. Conversely, elevated temperatures have potential to reduce the aromatic potential of grapes [52], which is possibly a reason for somewhat lower concentrations of monoterpenes in Dalmatian Pošip and Maraština wines. Kraljevina wines, which had the lowest concentrations of terpenes despite originating from the continental part, could be an exception that confirms the rule.

C_{13}-Norisoprenoids are also secondary metabolites in grapes, present in both aromatic and neutral varieties. They are formed as biodegradation products of carotenoid molecules, such as lutein, β-carotene, violaxanthin, and neoxanthin, via numerous formation mechanisms and intermediates during pre-fermentative steps, fermentation, and aging [53,54]. Four of them, β-damascenone, β-ionone, 1,1,6-trimethyl-1,2-dihydronaphthalene (TDN), and *trans*-1-(2,3,6-trimethylphenyl)buta-1,3-diene (TPB), were commonly found in wine at concentrations surpassing their odor perception thresholds, meaning they can have a direct impact on wine aroma [34]. Especially important is β-damascenone with its pleasant odor reminiscent of honey, dried plum and stewed apple, and a very low perception threshold, which ranks it among the most important wine odorants [37]. β-Ionone, characterized by a threshold of the similar order of magnitude, also significantly contributes to wine aroma with an odor reminiscent of violets, while the contribution of TDN and TPB becomes relevant mostly in aged wines [34]. The concentrations of the majority of C_{13}-norisoprenoids were generally higher in Dalmatian Pošip and Maraština, and the lowest in Kraljevina wines, although in particular cases with no statistical significance (Table 1). According to Marais and van Wyk [54] the concentration of β-damascenone is principally dependent on viticultural and winemaking conditions, while variety has less influence. Nevertheless, particular differences were observed: Malvazija wines were found to contain the highest concentration, although not different from that found in Pošip, while Škrlet had the lowest, not different from that found in Maraština wine. Malvazija was also characterized by the lowest concentration of vitispiranes together with Kraljevina wine. Among benzenoids, ethyl cinnamate emerged as a prominent marker of Pošip varietal origin, since it was found in the highest concentration in this wine.

C_6-alcohols are formed mainly in pre-fermentation vinification steps by degradation of unsaturated fatty acids by the action of enzymes, as well as by liberation from glycosidic precursors. They may have an effect on wine aroma with their so-called green and herbal odors, but luckily have relatively high odor perception thresholds, such as 8000 μg/L for 1-hexanol [37], so only very high concentration can produce negative effects. Certain authors include C_6-compounds among varietal aromas [16] and their concentrations were found useful in differentiation of particular wines based on variety [43,55]. The highest concentration of 1-hexanol was found in Škrlet, while Kraljevina contained the lowest amount (Table 1). Maraština, and especially Pošip wines were characterized by the highest concentration of unsaturated C_6-alcohols. It is possible that the mentioned differences were a consequence of different enzymatic potentials and fatty acid precursor loads in grapes of these varieties [55].

Concentrations and the composition of fermentation aroma compounds are mainly affected by fermentation conditions, but may also be influenced by grape composition [56]. Many studies proved that the composition of volatile compounds formed in fermentation can be useful in differentiating wines of mostly neutral varieties equally or even more successful than by using, e.g., monoterpene concentrations [11,14,20,29]. This is more characteristic for C_6–C_{10} fatty acids and the corresponding ethyl esters which, in contrast to acetates, are more dependent on the concentration of precursors and therefore on variety and conditions in vineyard, and less on the activity of yeast [57]. The average concentration of 2-phenylethanol was higher than the corresponding odor perception threshold of 10,000 µg/L in all the studied monovarietal wines, meaning this alcohol contributed significantly with its odor reminiscent of roses [37]. Pošip and Maraština had approximately 50% higher concentration of 2-phenylethanol in relation to the other investigated wines (Table 1). The concentrations of major volatile fatty acids (C_6–C_{10}) surpassed the corresponding odor perception thresholds of 420, 500, and 1000 µg/L, respectively [58], in all the investigated wines. Fatty acid production is determined in part by the initial composition of must [59] and therefore possibly by varietal origin. Malvazija istarska wines stood out with low concentrations of decanoic and octanoic acid. Among esters, Pošip was clearly differentiated from the other monovarietal wines by the highest concentration of 2-phenethyl acetate, which could have been related to the higher concentration of its precursor 2-phenylethanol found in this wine. However, it was stated previously that precursor concentrations do not significantly determine the concentrations of acetate esters formed by *Saccharomyces cerevisiae*, with the expression of alcohol acetyl transferase gene in yeast as a limiting factor [60]. Concentration of 2-phenethyl acetate in all the investigated wines was higher than the corresponding threshold of 250 µg/L [37], suggesting its floral odor participated in the aroma of all the wines. The major ethyl and acetate esters are among the most important volatile compounds for the fresh fruity aroma of young white wines to which they significantly contribute by commonly multiply surpassing their rather low odor perception thresholds, such as 30 µg/L for isoamyl acetate, 20 µg/L for ethyl butyrate, 5 µg/L for ethyl hexanoate, and 2 µg/L for ethyl octanoate [37]. The highest concentration of linear middle-chain ethyl esters and acetates other than 2-phenylethyl acetate, although in some cases without statistical significance, was noted in Malvazija istarska wines. Pošip was also relatively abundant in these esters, except for ethyl hexanoate which was found in the lowest concentration in this and in Maraština wines. Although hexanoic acid is mainly formed in fermentation, grapes also contain non-negligible concentration. This means that the concentration of ethyl hexanoate in wine is probably partly influenced by the concentration of its precursor, hexanoic acid, in grapes [4], so the lower concentration of ethyl hexanoate in Pošip and Maraština could have been influenced by a genotype.

3.2. GC×GC-TOF-MS

A characteristic HS-SPME/GC×GC-TOF-MS analysis 2D chromatogram of volatile compounds in Malvazija istarska wine is shown in Figure S1. It can be seen that many compounds which were separated by the second dimension column had the same retention times on the first, meaning these compounds would not be adequately separated by the conventional GC-MS. The average concentrations of volatile compounds (tentatively) identified in the investigated wines after GC×GC-TOF-MS analysis are reported in Table 2, while the concentrations found in each of the investigated wines are reported in Table S3. Compounds were grouped according to chemical class, and sorted within each class in order of decreasing *F*-ratio determined by one-way ANOVA. Three hundred and seventeen (317) volatile aroma compounds were identified, including 53 terpenes, 10 norisoprenoids, 50 benzenoids, 5 hydrocarbons, 7 aldehydes, 24 ketones, 32 alcohols, 16 acids, 73 esters, 5 volatile phenols, 17 furanoids and lactones, 19 sulfur containing compounds, and 6 other compounds. GC×GC-TOF-MS exhibited superior peak annotation ability than GC-MS which enabled the identification of a much larger number of compounds, as a consequence of higher separation efficiency, enhanced sensitivity, and clearer mass spectra allowed by separation on two different phases [23]. Other factors which could have affected the differences between the results obtained by the two techniques/methods were the absolute sensitivity of the analyzers, SPME conditions (sample volume and dilution, duration and temperature of extraction, fiber length, etc.), and others. To our knowledge, with 350 compounds identified by GC-MS and GC×GC-TOF-MS combined, this study reported one of the most detailed volatile aroma profiles in wine to date. It has to be noted that for particular compounds which were analyzed and reported by both the techniques applied the obtained absolute concentrations differed due to different quantification methods used: quantitative analysis with the use of standards solutions and calibration curves in GC-MS, and semi-quantification relative to internal standard 1-octanol concentration, assuming a response factor equal to one, in GC×GC-TOF-MS analysis, respectively. The concentrations of many volatile compounds were found to be significantly different between wines (161), but relatively few were found to be exclusive markers of particular variety.

In order to compare the techniques applied, the GC×GC-TOF-MS results for the major monoterpenols and some other compounds already quantified by GC-MS and reported in Table 1 were also reported in Table 2. It was observed that the results, in relative terms, were mostly in fair agreement. α-Terpineol was confirmed as a monoterpene and a volatile aroma compound in general with the highest discriminative power, with an *F*-ratio even higher than that obtained after GC-MS data elaboration. α-Terpineol was followed by limonene and linalool, as well as some other monoterpenes which were also among the most potent volatiles according to this criterion as determined by GC-MS, such as nerol, ho-trienol, 4-terpineol, and *trans*-β-ocimene. On the other hand, some discrepancies were observed; for example, in the case of geraniol, α-terpinolene, and geranyl ethyl ether, with a high *F*-ratio obtained by GC×GC-TOF-MS and a relatively low *F*-ratio obtained by GC-MS data elaboration. The opposite was observed for citronellol. It is possible that the discrepancies observed derived from the co-elution of the mentioned monoterpenes with particular unidentified compounds having mass spectra with ions of equal mass to those used for quantification of terpenes during GC-MS analysis, although strict measures have been taken to ensure the quality of the results.

Similar as in the case of GC-MS results (Table 1), Škrlet wines were the most abundant in monoterpenes, followed by Malvazija istarska, then Pošip, and finally Maraština and Kraljevina wines with the lowest concentrations (Table 2). Only a few exceptions were noted: Škrlet wines contained the lowest concentration of β-calacorene, while Malvazija wine was deficient in *cis*-Z-α-bisabolene epoxide. Although Kraljevina wine was generally poor in terpenes, several sesquiterpenes, such as cadalene, β-calacorene, and especially tentatively identified γ-dehydro-ar-himachalene, emerged as potential markers of the varietal origin of this wine.

All the other classes of compounds were confirmed to be far less efficient in differentiating the investigated monovarietal wines than terpenes, with few exceptions. The number of C_{13}-norisoprenoids identified by the two techniques applied was similar, but their identities differed in most cases. The relative results for β-damascenone obtained by GC-MS and GC×GC-TOF-MS were in a fair agreement, with the highest concentration found in Malvazija istarska and the lowest in Škrlet wines (Table 2). A similar degree of correspondence between GC-MS and GC×GC-TOF-MS results and the corresponding *F*-ratios was observed for a vitispirane isomer. Kraljevina wines contained the highest concentration of tentatively identified 1,2-dihydro-1,4,6-trimethylnaphthalene.

Superiority of GC×GC-TOF-MS over GC-MS in terms of compound separation and identification was demonstrated well in the analysis of benzenoids, with a much larger number of compounds identified by the former technique. Several benzenoids were found to be relatively efficient discriminators between monovarietal wines, and some of them were exclusive differentiators for particular varieties. High ethyl benzene concentration was specific for Pošip, while 1,1′-oxybisbenzene was most abundant in Malvazija istarska wines, in both cases supported by rather high *F*-ratios. In addition to the highest concentration of 1,1′-oxybisbenzene, Malvazija istarska wine was characterized by most varietal markers among benzenoids, including octylbenzene, a non-identified benzenoid, azulene, 2-methylnaphthalene, and methyl 2-(benzyloxy)propanoate. Pošip was characterized by the highest ethyl benzene and *trans*-edulan concentration, Kraljevina was the most abundant in 6-[1-(hydroxymethyl)vinyl]-4,8a-dimethyl-1,2,4a,5,6,7,8,8a-octahydro-2-naphthalenol, while Škrlet wine was the richest in *m*-methoxyanisole and α,α-dimethylbenzenemethanol (Table 2).

Table 2. Concentrations (μg/L relative to internal standard 2-octanol) of volatile aroma compounds found in Croatian monovarietal wines obtained by headspace solid-phase microextraction followed by comprehensive two-dimensional gas chromatography-mass spectrometry with time-of-flight mass spectrometric detection (HS-SPME/GC×GC-TOF-MS) sorted by compound class and descending Fisher *F*-ratio.

No.	Volatile Compounds	t_R (1D) (min:s)	t_R (2D) (min:s)	ID	LRI_{exp}	LRI_{lit}	*F*-Ratio	Variety				
								MI	PO	MA	KR	SK
	Terpenes											
1	α-Terpineol	18:47.2	00:01.2	MS, LRI	1710	1709	112.904	3.683 ± 1.391 [b]	2.440 ± 0.424 [c]	1.256 ± 0.455 [d]	0.712 ± 0.369 [d]	11.628 ± 0.495 [a]
2	Limonene	08:01.0	00:01.8	S, MS, LRI	1191	1194	54.231	1.401 ± 0.746 [b]	0.429 ± 0.261 [c]	0.360 ± 0.105 [c]	0.191 ± 0.114 [c]	3.956 ± 0.291 [a]
3	Linalool	15:50.0	00:01.1	S, MS, LRI	1541	1541	23.272	16.664 ± 7.072 [b]	10.757 ± 1.712 [c]	5.318 ± 1.567 [d]	2.857 ± 0.970 [d]	23.674 ± 3.494 [a]
4	*trans*-Alloocimene	12:13.0	00:01.6	MS, LRI	1384	1388	22.080	0.362 ± 0.189 [b]	0.137 ± 0.032 [c]	0.088 ± 0.044 [c]	0.040 ± 0.022 [c]	0.753 ± 0.282 [a]
5	*o*-Cymene	09:49.9	00:01.6	S, MS, LRI	1273	1268	20.745	1.305 ± 0.436 [b]	0.631 ± 0.353 [c]	0.441 ± 0.268 [c]	0.278 ± 0.142 [c]	2.613 ± 1.042 [a]
6	γ-Dehydro-ar-himachalene	25:10.0	00:01.7	MS	2046	-	18.720	0.003 ± 0.003 [b]	0.003 ± 0.003 [b]	0.003 ± 0.002 [b]	0.016 ± 0.006 [a]	0.004 ± 0.004 [b]
7	β-Myrcene	07:18.8	00:01.6	S, MS, LRI	1159	1159	17.244	2.351 ± 1.293 [b]	0.885 ± 0.539 [c]	0.331 ± 0.070 [c]	0.183 ± 0.094 [c]	4.353 ± 1.916 [a]
8	Nerol	20:43.5	00:01.0	S, MS, LRI	1812	1811	16.944	0.568 ± 0.277 [b]	0.220 ± 0.118 [c]	0.225 ± 0.070 [c]	0.077 ± 0.038 [c]	0.810 ± 0.139 [a]
9	*cis*-α-Ocimene	09:24.8	00:01.6	MS, LRI	1254	1255	14.724	2.278 ± 0.937 [a]	0.754 ± 0.205 [b]	0.459 ± 0.087 [b]	0.337 ± 0.170 [b]	3.070 ± 1.903 [a]
10	*trans*-Linalool furan oxide	13:39.9	00:01.2	S, MS, LRI	1445	1450	14.601	0.444 ± 0.117 [b]	0.480 ± 0.160 [b]	0.221 ± 0.144 [c]	0.157 ± 0.138 [c]	0.872 ± 0.264 [a]
11	Ho-trienol	17:07.0	00:01.0	MS, LRI	1607	1612	13.945	2.843 ± 1.206 [a]	2.307 ± 0.568 [a]	0.669 ± 0.405 [b]	0.712 ± 0.395 [b]	2.510 ± 0.219 [a]
12	Geraniol	21:33.0	00:01.0	S, MS, LRI	1856	1857	11.761	0.432 ± 0.254 [a]	0.225 ± 0.091 [b]	0.124 ± 0.041 [b]	0.084 ± 0.020 [b]	0.589 ± 0.079 [a]
13	Carvomenthenal	17:14.0	00:01.5	MS, LRI	1615	1629	11.321	0.381 ± 0.215 [a]	0.193 ± 0.088 [b]	0.142 ± 0.052 [b]	0.057 ± 0.034 [b]	0.525 ± 0.132 [a]
14	Linalool ethyl ether	11:02.8	00:01.9	MS, LRI	1329	1331	10.900	5.419 ± 2.787 [a]	1.744 ± 0.925 [b]	0.998 ± 0.204 [b]	0.444 ± 0.219 [b]	6.365 ± 4.476 [a]
15	4-Terpineol	17:00.0	00:01.3	S, MS, LRI	1600	1597	10.895	0.464 ± 0.097 [b]	0.363 ± 0.069 [cd]	0.418 ± 0.116 [bc]	0.289 ± 0.083 [d]	0.676 ± 0.045 [a]
16	α-Terpinolene	10:03.7	00:01.9	S, MS, LRI	1284	1282	10.846	1.630 ± 0.872 [b]	0.520 ± 0.288 [c]	0.387 ± 0.217 [c]	0.173 ± 0.085 [c]	3.391 ± 2.459 [a]
17	Geranyl ethyl ether	15:08.0	00:01.8	MS, LRI	1506	1506	10.391	2.042 ± 0.965 [a]	0.710 ± 0.328 [b]	0.500 ± 0.186 [b]	0.222 ± 0.161 [b]	2.960 ± 2.199 [a]
18	Monoterpene (n.i.; *m/z* 93, 121, 136)	13:54.1	00:01.9	MS	1455	-	9.899	10.628 ± 4.686 [a]	4.420 ± 1.974 [b]	2.737 ± 1.576 [b]	1.104 ± 0.702 [b]	16.227 ± 12.511 [a]
19	Neryl ethyl ether	14:26.5	00:01.8	MS, LRI	1477	1468	8.918	0.334 ± 0.175 [b]	0.143 ± 0.065 [bc]	0.100 ± 0.067 [bc]	0.023 ± 0.021 [c]	0.863 ± 0.729 [a]
20	Monoterpene (n.i.; *m/z* 68, 93, 121)	08:15.2	00:01.8	MS	1202	-	8.901	0.252 ± 0.272 [b]	0.126 ± 0.077 [bc]	0.047± 0.021 [c]	0.020 ± 0.015 [c]	0.575 ± 0.180 [a]
21	*p*-Cymenene	13:30.2	00:01.5	MS, LRI	1439	1438	7.409	1.332 ± 0.236 [a]	0.899 ± 0.190 [b]	0.918 ± 0.307 [b]	0.733 ± 0.210 [b]	1.711 ± 0.786 [a]
22	*trans*-β-Ocimene	09:01.7	00:01.7	S, MS, LRI	1237	1241	6.833	0.078 ± 0.148 [b]	0.067 ± 0.087 [b]	0.099 ± 0.109 [b]	0.014 ± 0.019 [b]	0.919 ± 0.926 [a]
23	β-Calacorene	22:43.0	00:01.8	MS, LRI	1919	1918	5.495	0.108 ± 0.018 [ab]	0.075 ± 0.011 [c]	0.078 ± 0.011 [c]	0.131 ± 0.044 [a]	0.072 ± 0.052 [c]
24	*cis*-Z-α-Bisabolene epoxide	24:28.0	00:01.3	MS, LRI	2010	2007	5.329	0.002 ± 0.003 [c]	0.008 ± 0.004 [bc]	0.010 ± 0.009 [b]	0.013 ± 0.006 [ab]	0.023 ± 0.016 [a]
25	*trans*-Linalool pyran oxide	19:34.0	00:01.1	MS, LRI	1751	1752	4.346	0.059 ± 0.018 [ab]	0.082 ± 0.049 [a]	0.029 ± 0.031 [b]	0.031 ± 0.019 [b]	0.086 ± 0.025 [a]
26	Cadalene	27:46.2	00:01.6	MS, LRI	>2100	2191	3.956	0.047 ± 0.010 [b]	0.039 ± 0.017 [b]	0.035 ± 0.011 [b]	0.064 ± 0.022 [a]	0.032 ± 0.021 [b]
27	Isogeraniol	20:55.4	00:01.0	MS, LRI	1822	1828	3.457	0.036 ± 0.030 [a]	0.018 ± 0.014 [ab]	0.005 ± 0.006 [b]	0.010 ± 0.011 [b]	0.015 ± 0.007 [ab]
28	α-Terpinene	07:33.6	00:01.8	S, MS, LRI	1170	1175	3.225	0.033 ± 0.061 [b]	0.006 ± 0.010 [b]	0.006 ± 0.010 [b]	0.007 ± 0.015 [b]	0.106 ± 0.120 [a]
29	Sesquiterpene (n.i.; *m/z* 119, 93, 69)	19:48.0	00:01.8	MS	1763	-	2.809	0.064 ± 0.010 [a]	0.038 ± 0.020 [b]	0.040 ± 0.014 [b]	0.045 ± 0.012 [b]	0.042 ± 0.038 [b]
30	Menthol	17:42.0	00:01.1	MS, LRI	1644	1641	2.746	0.129 ± 0.050 [b]	0.222 ± 0.283 [b]	0.115 ± 0.032 [b]	0.099 ± 0.023 [b]	1.177 ± 1.847 [a]
31	Citronellol	20:02.9	00:01.1	S, MS, LRI	1776	1777	2.724	0.303 ± 0.037 [ab]	0.250 ± 0.089 [ab]	0.260 ± 0.089 [ab]	0.154 ± 0.074 [b]	0.309 ± 0.229 [ab]

Table 2. *Cont.*

No.	Volatile Compounds	t_R (1D) (min:s)	t_R (2D) (min:s)	ID	LRI_{exp}	LRI_{lit}	*F*-Ratio	Variety MI	PO	MA	KR	SK
32	α-Farnesene II	19:48.0	00:01.9	S, MS, LRI	1763	1762	2.118	0.053 ± 0.018	0.029 ± 0.016	0.026 ± 0.023	0.039 ± 0.021	0.047 ± 0.026
33	Monoterpene (n.i.; *m/z* 93, 121, 94)	20:48.1	00:01.7	MS	1816	-	2.054	0.056 ± 0.034	0.067 ± 0.027	0.041 ± 0.041	0.028 ± 0.015	0.028 ± 0.009
34	γ-Cadinene	19:55.0	00:02.1	MS, LRI	1769	1774	1.895	0.020 ± 0.003	0.013 ± 0.007	0.016 ± 0.004	0.014 ± 0.007	0.012 ± 0.011
35	(+)-Cuparene	21:05.0	00:02.0	MS, LRI	1831	1830	1.847	0.020 ± 0.001	0.014 ± 0.010	0.017 ± 0.004	0.012 ± 0.007	0.012 ± 0.010
36	*trans*-Calamenene	21:19.0	00:01.9	MS, LRI	1844	1837	1.780	0.063 ± 0.019	0.056 ± 0.020	0.049 ± 0.011	0.077 ± 0.028	0.056 ± 0.023
37	Citronellyl acetate	18:10.0	00:01.5	MS, LRI	1673	1668	1.652	0.056 ± 0.031	0.082 ± 0.059	0.041 ± 0.022	0.035 ± 0.017	0.063 ± 0.060
38	α-Calacorene	22:29.0	00:01.8	MS, LRI	1906	1916	1.533	0.020 ± 0.003	0.014 ± 0.008	0.015 ± 0.005	0.019 ± 0.009	0.012 ± 0.010
39	Dehydroaromadendrene	28:20.9	00:01.4	MS	>2100	-	1.306	0.001 ± 0.002	0.014 ± 0.036	0.001 ± 0.002	0.012 ± 0.015	0.028 ± 0.034
40	Terpene (n.i.; *m/z* 121, 93, 136)	15:08.0	00:01.8	MS	1506	-	1.276	0.891 ± 0.202	0.784 ± 0.328	0.788 ± 0.357	0.458 ± 0.172	0.836 ± 1.082
41	β-Cyclocitral	17:21.0	00:01.5	S, MS, LRI	1622	1629	1.195	0.067 ± 0.016	0.071 ± 0.017	0.066 ± 0.030	0.051 ± 0.014	0.052 ± 0.028
42	α-Farnesene I	18:17.0	00:01.9	S, MS, LRI	1681	1697	1.124	0.126 ± 0.043	0.074 ± 0.034	0.102 ± 0.058	0.115 ± 0.061	0.113 ± 0.046
43	2-Acetyl-2-carene	23:11.0	00:01.3	MS	1943	-	1.088	0.030 ± 0.048	0.037 ± 0.039	0.047 ± 0.050	0.010 ± 0.013	0.054 ± 0.013
44	γ-Isogeraniol	20:29.2	00:01.0	MS, LRI	1799	1800	1.071	0.148 ± 0.134	0.123 ± 0.083	0.087 ± 0.064	0.061 ± 0.046	0.087 ± 0.078
45	Cosmene	13:41.6	00:01.4	MS, LRI	1446	1460	0.990	0.109 ± 0.117	0.056 ± 0.056	0.106 ± 0.068	0.080 ± 0.054	0.027 ± 0.011
46	α-Curcumene	20:16.0	00:01.8	MS, LRI	1787	1782	0.736	0.026 ± 0.014	0.019 ± 0.007	0.021 ± 0.007	0.029 ± 0.017	0.022 ± 0.011
47	*trans*-Geranyl acetone	21:40.0	00:01.5	MS, LRI	1863	1868	0.734	0.409 ± 0.437	0.228 ± 0.071	0.206 ± 0.060	0.422 ± 0.568	0.152 ± 0.036
48	α-Bergamotene	16:18.0	00:02.4	MS, LRI	1565	1585	0.675	0.027 ± 0.011	0.037 ± 0.027	0.058 ± 0.071	0.041 ± 0.020	0.043 ± 0.024
49	Sesquiterpene (n.i.; *m/z* 93, 80, 121)	20:16.0	00:02.0	MS	1787	-	0.596	0.021 ± 0.007	0.014 ± 0.008	0.020 ± 0.013	0.022 ± 0.014	0.022 ± 0.021
50	Neryl acetate	19:20.7	00:01.5	MS, LRI	1739	1742	0.576	0.085 ± 0.021	0.060 ± 0.050	0.058 ± 0.038	0.078 ± 0.054	0.092 ± 0.083
51	4-Thujanol	15:20.1	00:01.7	MS	1516	-	0.444	0.055 ± 0.040	0.051 ± 0.029	0.067 ± 0.043	0.068 ± 0.024	0.046 ± 0.012
52	Nerolidol	24:56.2	00:01.3	S, MS, LRI	2034	2034	0.434	0.114 ± 0.049	0.102 ± 0.040	0.124 ± 0.087	0.142 ± 0.058	0.131 ± 0.064
53	Geranyl acetate	19:55.0	00:01.5	S, MS, LRI	1769	1768	0.413	0.037 ± 0.015	0.048 ± 0.038	0.035 ± 0.018	0.034 ± 0.013	0.034 ± 0.033
	C_{13}-norisoprenoids											
54	1,2-Dihydro-1,4,6-trimethylnaphthalene	25:38.3	00:01.5	MS	2071	-	7.148	0.001 ± 0.001 ^{b}	0.000 ± 0.000 ^{b}	0.009 ± 0.008 ^{b}	0.023 ± 0.018 ^{a}	0.002 ± 0.001 ^{b}
55	β-Damascenone	21:05.0	00:01.5	S, MS, LRI	1831	1832	6.736	8.245 ± 2.169 ^{a}	5.770 ± 2.963 ^{bc}	4.338 ± 1.563 ^{cd}	6.981 ± 1.067 ^{ab}	2.336 ± 0.303 ^{d}
56	α-Ionene	14:28.9	00:02.0	MS	1479	-	5.379	0.045 ± 0.017 ^{bc}	0.017 ± 0.023 ^{c}	0.069 ± 0.047 ^{ab}	0.085 ± 0.031 ^{a}	0.032 ± 0.017 ^{bc}
57	Norisoprenoid (n.i.; *m/z* 69, 121, 105)	20:00.5	00:01.6	MS	1774	-	5.061	0.248 ± 0.076 ^{a}	0.163 ± 0.101 ^{b}	0.127 ± 0.060 ^{bc}	0.202 ± 0.058 ^{ab}	0.055 ± 0.025 ^{c}
58	3,4-Dehydro-β-ionone	18:38.0	00:01.8	MS	1702	-	4.920	0.032 ± 0.009 ^{a}	0.011 ± 0.010 ^{b}	0.036 ± 0.017 ^{a}	0.040 ± 0.020 ^{a}	0.020 ± 0.004 ^{ab}
59	Vitispirane I	15:29.0	00:01.9	MS, LRI	1524	1524	3.470	1.155 ± 0.481 ^{c}	3.501 ± 1.555 ^{ab}	4.058 ± 2.588 ^{a}	2.200 ± 0.966 ^{bc}	3.007 ± 2.552 ^{abc}
60	1,2-Dihydro-1,5,8-trimethylnaphthalene	19:41.0	00:01.6	MS	1757	-	2.579	0.360 ± 0.109	0.363 ± 0.186	0.682 ± 0.399	0.770 ± 0.451	1.085 ± 1.062
61	Actinidol ethyl ether II	18:52.0	00:01.9	MS, LRI	1714	1723	2.179	0.099 ± 0.082	0.181 ± 0.106	0.167 ± 0.130	0.055 ± 0.030	0.133 ± 0.067
62	1,2-Dihydro-1,1,6-trimethylnaphthalene (TDN)	19:06.0	00:01.6	S, MS, LRI	1727	1729	0.885	0.021 ± 0.012	0.019 ± 0.010	0.031 ± 0.021	0.026 ± 0.018	0.015 ± 0.015

Table 2. *Cont.*

No.	Volatile Compounds	t_R (1D) (min:s)	t_R (2D) (min:s)	ID	LRI_{exp}	LRI_{lit}	*F*-Ratio	Variety MI	PO	MA	KR	SK
63	*trans*-1-(2,3,6-Trimethylphenyl)buta-1,3-diene (TPB)	21:12.0	00:01.5	MS, LRI	1837	1832	0.433	0.065 ± 0.029	0.057 ± 0.036	0.076 ± 0.048	0.057 ± 0.031	0.050 ± 0.025
	Benzenoids											
64	Ethyl benzoate	18:17.0	00:01.2	MS, LRI	1681	1678	20.194	0.759 ± 0.232 b	1.493 ± 0.366 a	0.653 ± 0.128 b	0.533 ± 0.118 b	0.570 ± 0.115 b
65	1,1′-Oxybisbenzene	24:20.1	00:01.3	MS, LRI	2003	2017	18.956	0.011 ± 0.002 a	0.004 ± 0.002 b	0.004 ± 0.003 b	0.003 ± 0.001 b	0.003 ± 0.001 b
66	2,3-Dihydro-1,1,5,6-tetramethyl-1H-indene I	18:17.0	00:01.7	MS	1681	-	9.842	0.076 ± 0.019 bc	0.027 ± 0.030 d	0.105 ± 0.039 ab	0.139 ± 0.054 a	0.050 ± 0.013 cd
67	Octylbenzene	19:27.0	00:01.8	MS, LRI	1745	1746	8.638	0.109 ± 0.011 a	0.065 ± 0.022 b	0.074 ± 0.010 b	0.072 ± 0.012 b	0.070 ± 0.032 b
68	*trans*-Edulan	17:07.0	00:01.8	MS, LRI	1607	1602	7.938	0.039 ± 0.016 b	0.084 ± 0.018 a	0.056 ± 0.019 b	0.042 ± 0.010 b	0.035 ± 0.031 b
69	2,3-Dihydro-1,1,5,6-tetramethyl-1H-indene II	17:28.0	00:01.7	MS	1629	-	7.671	0.023 ± 0.006 bc	0.007 ± 0.010 c	0.038 ± 0.020 ab	0.056 ± 0.030 a	0.016 ± 0.004 bc
70	3-Methylphenylacetylene	14:33.0	00:01.3	MS, LRI	1481	1450.9	7.438	0.012 ± 0.002 b	0.016 ± 0.006 b	0.013 ± 0.007 b	0.032 ± 0.013 a	0.028 ± 0.012 a
71	Benzoic acid	30:48.9	00:00.8	S, MS, LRI	>2100	2438	6.952	0.269 ± 0.021 bc	0.412 ± 0.064 a	0.350 ± 0.047 a	0.233 ± 0.107 c	0.334 ± 0.111 ab
72	Azulene	19:34.0	00:01.3	MS, LRI	1751	1746	6.891	0.240 ± 0.058 a	0.181 ± 0.044 b	0.165 ± 0.042 bc	0.130 ± 0.039 c	0.117 ± 0.039 c
73	Trimethyl-tetra hydronaphthalene	13:12.1	00:01.8	MS	1426	-	6.830	0.007 ± 0.012 c	0.002 ± 0.004 c	0.056 ± 0.056 ab	0.102 ± 0.070 a	0.007 ± 0.008 c
74	Benzeneacetaldehyde	17:49.0	00:01.1	S, MS, LRI	1651	1648	6.827	5.920 ± 1.513 c	10.269 ± 2.174 a	8.487 ± 2.065 ab	5.884 ± 2.053 c	6.501 ± 1.630 bc
75	*m*-Methoxyanisole	19:48.0	00:01.2	MS, LRI	1763	1761	6.715	0.009 ± 0.013 c	0.035 ± 0.021 b	0.009 ± 0.011 c	0.013 ± 0.013 bc	0.070 ± 0.052 a
76	Benzenoid (n.i.; *m/z* 115, 130, 129)	16:37.2	00:01.4	MS	1581	-	5.764	0.007 ± 0.003 b	0.005 ± 0.004 b	0.005 ± 0.004 b	0.014 ± 0.005 a	0.009 ± 0.003 ab
77	6-[1-(Hydroxymethyl)vinyl] -4,8a-dimethyl-1,2,4a,5,6,7,8, 8a-octahydro-2-naphthalenol	19:54.7	00:02.0	MS	1769	-	5.587	0.010 ± 0.014 b	0.010 ± 0.011 b	0.009 ± 0.010 b	0.041 ± 0.023 a	0.013 ± 0.011 b
78	Prehnitene	14:40.0	00:01.5	MS, LRI	1486	1476	5.516	0.198 ± 0.035 ab	0.160 ± 0.052 bc	0.249 ± 0.080 a	0.124 ± 0.034 c	0.124 ± 0.089 c
79	2-Methylnaphthalene	22:15.0	00:01.3	MS, LRI	1894	1872	4.321	0.021 ± 0.004 a	0.015 ± 0.003 b	0.014 ± 0.004 b	0.015 ± 0.005 b	0.012 ± 0.003 b
80	*meso*-2,3-Diphenylbutane	17:07.0	00:01.3	MS	1607	-	4.291	0.025 ± 0.017 b	0.050 ± 0.016 a	0.037 ± 0.009 ab	0.027 ± 0.010 b	0.032 ± 0.007 ab
81	4-Ethylbenzaldehyde	19:34.0	00:01.2	MS, LRI	1751	1747	4.168	0.059 ± 0.010 ab	0.067 ± 0.014 a	0.060 ± 0.015 ab	0.043 ± 0.010 c	0.048 ± 0.008 bc
82	Styrene	09:27.4	00:05.0	MS, LRI	1256	1257	3.578	2.067 ± 0.516 ab	2.462 ± 0.859 a	2.161 ± 0.275 ab	1.701 ± 0.419 bc	1.223 ± 0.083 c
83	Ethyl *o*-methylbenzoate	19:34.2	00:01.3	MS	1751	-	3.148	0.039 ± 0.005 ab	0.041 ± 0.005 a	0.034 ± 0.005 abc	0.028 ± 0.013 c	0.027 ± 0.015 c
84	2,3-Dihydrobenzofuran	16:46.0	00:01.2	MS	1588	-	3.122	0.025 ± 0.011 b	0.046 ± 0.014 a	0.039 ± 0.013 ab	0.029 ± 0.014 b	0.029 ± 0.001 b
85	Benzofuran	15:01.0	00:01.1	MS, LRI	1500	1496	3.121	0.040 ± 0.013 b	0.061 ± 0.022 a	0.047 ± 0.015 ab	0.040 ± 0.013 b	0.030 ± 0.005 b
86	Benzonitrile	17:00.0	00:01.0	MS, LRI	1600	1591	3.041	0.033 ± 0.015 b	0.064 ± 0.026 a	0.052 ± 0.021 ab	0.038 ± 0.016 b	0.039 ± 0.005 ab
87	α,α-Dimethylbenzene methanol	19:55.2	00:01.0	MS, LRI	1769	1770	2.981	0.021 ± 0.009 b	0.033 ± 0.018 b	0.024 ± 0.008 b	0.027 ± 0.007 b	0.203 ± 0.309 a
88	Methyl 2-(benzyloxy)propanoate	23:18.0	00:01.3	MS	1949	-	2.958	0.825 ± 0.861 a	0.144 ± 0.151 b	0.107 ± 0.138 b	0.211 ± 0.456 b	0.018 ± 0.017 b
89	α-Methylstyrene	11:10.8	00:01.3	MS, LRI	1336	1325	2.727	0.006 ± 0.007	0.014 ± 0.006	0.015 ± 0.007	0.017 ± 0.008	0.118 ± 0.192
90	3-Ethylbenzaldehyde	18:59.4	00:01.2	MS, LRI	1721	1732	2.631	0.088 ± 0.013	0.092 ± 0.016	0.086 ± 0.023	0.065 ± 0.021	0.075 ± 0.004
91	3-Methylbenzofuran	21:19.0	00:01.1	MS	1844	-	2.594	0.027 ± 0.006	0.046 ± 0.012	0.034 ± 0.016	0.032 ± 0.014	0.034 ± 0.006
92	Styralyl isobutyrate	23:31.8	00:01.4	MS	1961	-	2.462	0.065 ± 0.022	0.191 ± 0.121	0.166 ± 0.115	0.149 ± 0.069	0.097 ± 0.023
93	2′,5′-Dimethylcrotonophenone	24:00.0	00:01.3	MS	1985	-	2.362	0.045 ± 0.017	0.040 ± 0.029	0.026 ± 0.019	0.040 ± 0.011	0.010 ± 0.009

Table 2. *Cont.*

No.	Volatile Compounds	t_R (1D) (min:s)	t_R (2D) (min:s)	ID	LRI_{exp}	LRI_{lit}	*F*-Ratio	Variety MI	PO	MA	KR	SK
94	Ethyl benzenepropanoate	22:15.0	00:01.3	MS, LRI	1894	1892	2.328	0.404 ± 0.256	0.509 ± 0.353	0.314 ± 0.180	0.206 ± 0.052	0.131 ± 0.079
95	1-Methylnapthalene	21:40.0	00:01.3	MS, LRI	1863	1878	2.319	0.017 ± 0.005	0.016 ± 0.007	0.014 ± 0.003	0.011 ± 0.003	0.010 ± 0.001
96	Ethyl salicylate	21:35.8	00:01.1	S, MS, LRI	1859	1837	2.288	0.037 ± 0.054	0.005 ± 0.008	0.027 ± 0.049	0.011 ± 0.012	0.177 ± 0.296
97	Methyl salicylate	20:16.0	00:01.2	S, MS, LRI	1787	1789	2.151	3.457 ± 1.576	4.548 ± 6.357	2.447 ± 1.435	1.979 ± 0.985	14.629 ± 21.656
98	2-Methylbenzaldehyde	17:27.8	00:01.1	MS, LRI	1629	1622	2.145	0.041 ± 0.014	0.057 ± 0.023	0.034 ± 0.023	0.032 ± 0.012	0.051 ± 0.025
99	Durene	13:16.0	00:01.5	MS, LRI	1429	1435	2.038	0.085 ± 0.033	0.084 ± 0.040	0.080 ± 0.027	0.056 ± 0.034	0.034 ± 0.028
100	Butylated hydroxytoluene	22:43.0	00:01.5	MS, LRI	1919	1920	1.558	0.314 ± 0.073	0.351 ± 0.163	0.295 ± 0.107	0.228 ± 0.067	0.210 ± 0.146
101	Ethyl benzeneacetate	20:30.0	00:01.2	MS, LRI	1799	1788	1.474	1.334 ± 0.310	3.173 ± 0.863	2.503 ± 1.245	3.294 ± 3.186	3.112 ± 2.566
102	*p*-Methoxyanisole	19:34.2	00:01.2	MS, LRI	1751	1752	1.309	0.153 ± 0.041	0.184 ± 0.074	0.141 ± 0.044	0.197 ± 0.052	0.191 ± 0.075
103	Benzeneacetic acid	32:45.2	00:00.8	MS, LRI	>2100	2519	1.243	0.002 ± 0.000	0.005 ± 0.005	0.008 ± 0.007	0.004 ± 0.007	0.009 ± 0.013
104	Benzyl alcohol	22:01.0	00:00.9	S, MS, LRI	1881	1877	1.243	0.914 ± 1.413	2.007 ± 2.953	0.468 ± 0.243	0.354 ± 0.079	0.565 ± 0.297
105	2-Hydroxybenzeneacetic acid	23:11.0	00:01.0	MS	1943	-	1.238	0.003 ± 0.008	0.008 ± 0.004	0.003 ± 0.003	0.005 ± 0.001	0.005 ± 0.002
106	2-Ethyl-*m*-xylene	11:58.5	00:01.6	MS, LRI	1373	1372	1.224	0.106 ± 0.064	0.112 ± 0.059	0.082 ± 0.043	0.094 ± 0.040	0.040 ± 0.035
107	Benzaldehyde	15:16.8	00:05.8	S, MS, LRI	1514	1509	1.094	1.935 ± 0.978	4.662 ± 6.475	1.479 ± 0.377	2.161 ± 0.535	2.934 ± 2.144
108	2-(1,1-Dimethylethyl)-1,4-dimethoxybenzene	22:29.0	00:01.4	MS, LRI	1906	1870	0.996	0.069 ± 0.010	0.061 ± 0.014	0.066 ± 0.017	0.075 ± 0.021	0.051 ± 0.044
109	*trans*-1,2-Diphenylcyclobutane	20:58.0	00:01.3	MS	1825	-	0.674	0.003 ± 0.002	0.003 ± 0.002	0.004 ± 0.007	0.002 ± 0.003	0.000 ± 0.000
110	3,3-Dimethoxy-1-phenyl propan-1-one	09:56.8	00:05.6	MS	1278	-	0.569	0.037 ± 0.032	0.048 ± 0.054	0.052 ± 0.059	0.025 ± 0.027	0.062 ± 0.031
111	*trans*-Anethole	21:12.0	00:01.3	S, MS, LRI	1837	1834	0.502	0.423 ± 0.223	0.449 ± 0.199	0.390 ± 0.213	0.395 ± 0.309	0.613 ± 0.344
112	*cis*-Anethole	19:55.0	00:01.3	S, MS, LRI	1769	1780	0.322	0.013 ± 0.005	0.015 ± 0.007	0.014 ± 0.004	0.014 ± 0.007	0.017 ± 0.010
113	1,2-Dimethylbenzene	07:41.0	00:05.1	MS, LRI	1176	1175	0.176	0.468 ± 0.333	0.412 ± 0.402	0.503 ± 0.355	0.356 ± 0.312	0.432 ± 0.375
	Hydrocarbons											
114	Pentadecane	14:54.0	00:02.7	S, MS, LRI	1496	1500	1.699	0.251 ± 0.055	0.195 ± 0.066	0.222 ± 0.104	0.205 ± 0.034	0.140 ± 0.045
115	2,3,3-Trimethyl-*cis*-4-nonene	10:42.0	00:01.6	MS	1313	-	1.585	0.078 ± 0.095	0.052 ± 0.034	0.031 ± 0.008	0.016 ± 0.009	0.026 ± 0.031
116	Hexadecane	17:00.0	00:02.7	S, MS, LRI	1600	1600	1.332	0.170 ± 0.025	0.140 ± 0.051	0.162 ± 0.082	0.134 ± 0.038	0.100 ± 0.031
117	*cis,trans*-1,3,5-Octatriene	08:08.4	00:00.7	MS	1196	-	1.087	0.376 ± 0.146	0.364 ± 0.125	0.315 ± 0.140	0.245 ± 0.137	0.329 ± 0.029
118	2,6,8-Trimethyl-*trans*-4-nonene	11:13.5	00:02.5	MS	1338	-	0.464	0.048 ± 0.083	0.029 ± 0.052	0.019 ± 0.039	0.032 ± 0.043	0.002 ± 0.002
	Aldehydes											
119	Decanal	14:54.0	00:01.5	S, MS, LRI	1496	1497	3.149	0.068 ± 0.041 [a]	0.068 ± 0.035 [a]	0.051 ± 0.040 [ab]	0.014 ±0.011 [b]	0.060 ± 0.008 [ab]
120	*trans*-2-Decenal	17:49.0	00:01.4	MS, LRI	1651	1647	2.553	0.116 ± 0.030	0.131 ± 0.043	0.109 ± 0.030	0.071 ± 0.034	0.094 ± 0.065
121	*trans*-2-Octenal	13:25.3	00:01.3	S, MS, LRI	1435	1432	2.307	0.249 ± 0.301	0.017 ± 0.023	0.120 ± 0.038	0.105 ± 0.092	0.021 ± 0.020
122	Undecanal	17:07.0	00:01.5	S, MS, LRI	1607	1606	1.967	0.061 ± 0.020	0.053 ± 0.016	0.040 ± 0.010	0.041 ± 0.010	0.051 ± 0.031
123	Dodecanal	19:06.0	00:01.6	MS, LRI	1727	1722	1.471	0.066 ± 0.015	0.062 ± 0.019	0.050 ± 0.014	0.056 ± 0.022	0.043 ± 0.011
124	3,3-Dimethyl-2-oxobutanal	11:24.8	00:01.9	MS	1347	-	0.279	0.108 ± 0.091	0.158 ± 0.178	0.106 ± 0.104	0.226 ± 0.522	0.078 ± 0.082
125	Nonanal	12:40.6	00:01.5	S, MS, LRI	1405	1404	0.206	10.699 ± 7.689	12.183 ± 7.410	11.470 ± 8.481	8.722 ± 6.668	10.800 ± 6.868
	Ketones											
126	1,4,7,10,13-Pentaoxa cyclononadecane-14,19-dione	27:35.0	00:01.3	MS	>2100	-	9.721	0.027 ± 0.038 [b]	0.013 ± 0.014 [b]	0.005 ± 0.003 [b]	0.022 ± 0.020 [b]	0.107 ± 0.038 [a]
127	α-Isophorone	16:46.0	00:01.3	S, MS, LRI	1588	1593	7.380	0.116 ± 0.022 [a]	0.101 ± 0.032 [a]	0.047 ± 0.024 [b]	0.093 ± 0.021 [a]	0.093 ± 0.029 [a]

Table 2. *Cont.*

No.	Volatile Compounds	t_R (1D) (min:s)	t_R (2D) (min:s)	ID	LRI_{exp}	LRI_{lit}	*F*-Ratio	Variety				
								MI	PO	MA	KR	SK
128	Cyclohexylideneacetone	17:35.0	00:01.8	MS	1637	-	6.967	0.097 ± 0.097 [c]	0.671 ± 0.428 [b]	0.424 ± 0.259 [bc]	0.546 ± 0.254 [b]	1.194 ± 0.691 [a]
129	Acetophenone	17:56.0	00:01.1	S, MS, LRI	1659	1660	6.036	0.297 ± 0.048 [cd]	0.557 ± 0.116 [a]	0.415 ± 0.107 [bc]	0.271 ± 0.078 [d]	0.487 ± 0.348 [ab]
130	2-Undecanone	16:53.4	00:01.5	MS, LRI	1594	1598	4.027	0.755 ± 0.471 [a]	0.320 ± 0.097 [b]	0.294 ± 0.104 [b]	0.347 ± 0.246 [b]	0.215 ± 0.162 [b]
131	4,4-(Ethylenedioxy)-2-pentanone	19:20.0	00:01.1	MS	1739	-	3.787	0.166 ± 0.053 [ab]	0.245 ± 0.111 [a]	0.145 ± 0.089 [b]	0.102 ± 0.020 [b]	0.102 ± 0.067 [b]
132	Unsaturated diketone (n.i.; *m/z* 43, 99, 71)	14:26.0	00:01.1	MS	1477	-	3.213	0.137 ± 0.151 [b]	0.361 ± 0.123 [a]	0.269 ± 0.095 [ab]	0.170 ± 0.174 [b]	0.150 ± 0.114 [b]
133	3-Undecanone	16:20.3	00:01.6	MS, LRI	1567	1571	3.176	0.455 ± 0.440 [a]	0.075 ± 0.051 [b]	0.081 ± 0.040 [b]	0.123 ± 0.265 [b]	0.025 ± 0.021 [b]
134	3-Tridecanone	20:16.0	00:01.7	MS, LRI	1787	1755	3.120	0.036 ± 0.023 [a]	0.010 ± 0.008 [b]	0.008 ± 0.009 [b]	0.018 ± 0.026 [ab]	0.006 ± 0.005 [b]
135	1b,5,5,6a-Tetramethyl-octahydro-1-oxa-cyclopropa [a]inden-6-one	18:31.0	00:02.1	MS	1695	-	2.570	0.016 ± 0.019	0.042 ± 0.038	0.047 ± 0.054	0.088 ± 0.068	0.123 ± 0.138
136	3-(Acetoxy)-4-methyl-2-pentanone	14:07.3	00:01.3	MS	1464	-	2.528	0.012 ± 0.021	0.055 ± 0.055	0.073 ± 0.057	0.025 ± 0.025	0.031 ± 0.027
137	*trans*-5-Methyl-2-(1-methylethyl)-cyclohexanone	14:08.1	00:01.6	MS, LRI	1464	1473	2.282	0.041 ± 0.045	0.312 ± 0.657	0.077 ± 0.050	0.058 ± 0.030	1.167 ± 1.900
138	4-(1,1-Dimethylethyl)-cyclohexanone	17:35.0	00:01.5	MS, LRI	1637	1645	2.032	0.057 ± 0.095	0.124 ± 0.122	0.025 ± 0.033	0.011 ± 0.011	0.092 ± 0.125
139	1-Phenyl-1-propanone	19:20.2	00:01.2	MS, LRI	1739	1744	1.745	0.019 ± 0.018	0.030 ± 0.013	0.015 ± 0.007	0.015 ± 0.004	0.021 ± 0.005
140	2-Nonanone	12:34.0	00:01.4	S, MS, LRI	1401	1402	1.689	10.162 ± 9.346	4.449 ± 3.737	6.475 ± 7.123	3.195 ± 1.103	2.225 ± 2.037
141	2H-Pyran-2,6(3H)-dione	24:10.5	00:00.8	MS	1995	-	1.599	0.485 ± 0.190	0.661 ± 0.326	0.564 ± 0.226	0.397 ± 0.138	0.394 ± 0.069
142	2-Heptanone	07:46.5	00:05.2	S, MS, LRI	1180	1180	1.445	0.692 ± 0.406	0.605 ± 0.416	1.117 ± 1.197	0.301 ± 0.265	0.502 ± 0.401
143	2,2-Dimethyl-1,3-dioxane-4,6-dione	18:17.0	00:01.1	MS	1681	-	1.167	0.032 ± 0.014	0.022 ± 0.021	0.037 ± 0.010	0.035 ± 0.006	0.026 ± 0.023
144	2-Decanone	14:47.0	00:01.5	MS, LRI	1491	1491	0.990	0.502 ± 0.298	0.457 ± 0.128	0.377 ± 0.175	0.432 ± 0.159	0.255 ± 0.107
145	Acetoin	10:07.9	00:00.8	S, MS, LRI	1287	1287	0.732	0.054 ± 0.017	0.071 ± 0.048	0.061 ± 0.026	0.093 ± 0.083	0.058 ± 0.024
146	2,6-Di(tert-butyl)-4-hydroxy-4-methyl-2,5-cyclohexadien-1-one	25:45.0	00:01.1	MS, LRI	2077	2094	0.694	0.009 ± 0.003	0.010 ± 0.006	0.010 ± 0.007	0.014 ± 0.011	0.013 ± 0.008
147	2-Cyclohexene-1,4-dione	19:33.7	00:01.0	MS	1751	-	0.669	0.006 ± 0.010	0.027 ± 0.031	0.093 ± 0.214	0.031 ± 0.060	0.055 ± 0.078
148	3,4-Dihydroxy-cyclobutene-1,2-dione	18:10.0	00:01.1	MS	1673	-	0.409	0.089 ± 0.031	0.075 ± 0.037	0.060 ± 0.053	0.078 ± 0.059	0.073 ± 0.022
149	5-Methyl-5-hepten-2-one	11:24.0	00:01.3	MS, LRI	1346	1343	0.215	0.147 ± 0.160	0.113 ± 0.085	0.125 ± 0.023	0.116 ± 0.112	0.170 ± 0.130
	Alcohols											
150	4-Methyl-1-heptanol	12:42.1	00:01.6	MS, LRI	1406	1409	23.056	0.313 ± 0.161 [b]	0.115 ± 0.060 [c]	0.056 ± 0.047 [c]	0.033 ± 0.031 [c]	0.638 ± 0.209 [a]
151	*cis*-3-Hexen-1-ol	12:20.0	00:00.9	S, MS, LRI	1390	1386	15.611	7.191 ± 2.621 [c]	15.988 ± 3.409 [a]	11.216 ± 4.880 [b]	3.383 ± 0.607 [d]	5.727 ± 3.198 [cd]
152	2-Heptanol	10:56.0	00:01.0	S, MS, LRI	1324	1320	7.290	0.943 ± 0.327 [bc]	1.984 ± 0.923 [a]	1.044 ± 0.321 [bc]	0.601 ± 0.251 [c]	1.571 ± 0.480 [ab]
153	2-Penten-1-ol	10:56.2	00:00.8	MS, LRI	1324	1321	5.588	0.044 ± 0.021 [a]	0.045 ± 0.014 [a]	0.044 ± 0.017 [a]	0.012 ± 0.002 [b]	0.040 ± 0.012 [a]
154	3-Octanol	12:36.4	00:01.1	MS, LRI	1402	1406	5.108	0.082 ± 0.068 [b]	0.185 ± 0.070 [a]	0.062 ± 0.053 [b]	0.072 ± 0.050 [b]	0.054 ± 0.054 [b]
155	1-Undecanol	21:49.4	00:01.1	MS, LRI	1871	1883	5.052	0.004 ± 0.006 [b]	0.023 ± 0.015 [a]	0.007 ± 0.012 [b]	0.022 ± 0.010 [a]	0.022 ± 0.007 [a]
156	Alcohol (n.i.; *m/z* 69, 41, 84)	15:02.3	00:01.0	MS	1501	-	4.278	0.491 ± 0.778 [bc]	0.848 ± 0.393 [ab]	1.178 ± 0.637 [a]	0.192 ± 0.081 [c]	0.079 ± 0.036 [c]
157	1-Octen-3-ol	13:51.0	00:01.0	S, MS, LRI	1453	1452	3.832	3.494 ± 2.869 [b]	5.797 ± 1.668 [a]	2.467 ± 0.654 [b]	2.597 ± 1.254 [b]	3.015 ± 1.251 [b]

Table 2. *Cont.*

No.	Volatile Compounds	t_R (1D) (min:s)	t_R (2D) (min:s)	ID	LRI_{exp}	LRI_{lit}	*F*-Ratio	Variety				
								MI	PO	MA	KR	SK
158	2-Decanol	17:20.3	00:01.1	MS, LRI	1621	1621	3.058	0.023 ± 0.011 [b]	0.047 ± 0.025 [b]	0.021 ± 0.015 [b]	0.024 ± 0.013 [b]	0.212 ± 0.318 [a]
159	2,3-Butanediol II	16:25.2	00:00.8	S, MS, LRI	1571	1567	2.708	1.964 ± 0.450	2.601 ± 0.580	2.586 ± 1.276	1.421 ± 0.780	2.007 ± 0.180
160	4-Hepten-1-ol	14:54.5	00:00.9	MS, LRI	1496	1502	2.635	0.105 ± 0.065	0.093 ± 0.090	0.125 ± 0.060	0.044 ± 0.025	0.172 ± 0.057
161	6-Methyl-5-hepten-2-ol	14:05.8	00:01.0	S, MS, LRI	1463	1466	2.465	0.034 ± 0.006	0.050 ± 0.015	0.042 ± 0.013	0.037 ± 0.011	0.030 ± 0.012
162	3-Methyl-1-pentanol	11:03.7	00:00.9	S, MS, LRI	1330	1332	2.387	5.790 ± 1.421	5.799 ± 1.504	6.404 ± 2.937	3.575 ± 1.820	4.453 ± 0.203
163	2,3-Butanediol I	15:36.0	00:02.1	S, MS, LRI	1530	1542	2.291	3.399 ± 1.779	4.339 ± 1.812	3.931 ± 2.279	1.744 ± 1.027	3.597 ± 0.875
164	Alcohol (n.i.; *m/z* 45, 55, 43)	14:17.6	00:01.0	MS	1471	-	2.157	0.137 ± 0.120	0.012 ± 0.022	0.096 ± 0.114	0.093 ± 0.042	0.054 ± 0.043
165	1-Decanol	20:02.0	00:01.1	S, MS, LRI	1775	1778	2.086	0.710 ± 0.219	0.625 ± 0.100	0.597 ± 0.138	0.812 ± 0.097	0.642 ± 0.209
166	3,5-Dimethyl-4-heptanol	19:35.2	00:00.8	MS	1752	-	2.033	0.043 ± 0.031	0.104 ± 0.063	0.098 ± 0.058	0.053 ± 0.056	0.068 ± 0.002
167	2-Ethylhexanol	14:40.0	00:01.0	MS, LRI	1486	1484	1.756	4.569 ± 0.735	4.664 ± 0.537	4.826 ± 0.481	4.169 ± 0.465	4.122 ± 0.201
168	8-Methyl-1,8-nonanediol	10:56.2	00:01.1	MS	1324	-	1.454	0.123 ± 0.079	0.199 ± 0.084	0.223 ± 0.122	0.206 ± 0.084	0.133 ± 0.116
169	3,4-Nonadienol	19:48.0	00:01.0	MS, LRI	1763	1754	1.445	0.017 ± 0.025	0.016 ± 0.013	0.007 ± 0.005	0.003 ± 0.002	0.005 ± 0.001
170	1-Pentanol	08:53.5	00:00.5	S, MS, LRI	1231	1242	1.272	0.082 ± 0.091	0.135 ± 0.100	0.112 ± 0.076	0.058 ± 0.060	0.039 ± 0.014
171	3-Ethyl-4-octanol	18:24.9	00:01.3	MS	1689	-	1.225	0.364 ± 0.208	0.304 ± 0.189	0.479 ± 0.198	0.471 ± 0.108	0.331 ± 0.229
172	2-Octen-1-ol	17:14.0	00:01.0	S, MS, LRI	1615	1622	1.115	0.151 ± 0.206	0.114 ± 0.060	0.077 ± 0.033	0.041 ± 0.015	0.057 ± 0.020
173	2-Undecanol	19:13.0	00:01.2	MS, LRI	1733	1738	0.841	0.173 ± 0.218	0.330 ± 0.184	0.226 ± 0.137	0.241 ± 0.094	0.278 ± 0.205
174	4-Methyl-1-pentanol	10:49.0	00:00.9	MS, LRI	1319	1319	0.819	2.297 ± 1.083	2.626 ± 0.800	3.318 ± 2.010	2.130 ± 2.101	1.782 ± 0.130
175	4-Ethyl-3-octanol	15:08.0	00:01.2	MS	1506	-	0.721	0.400 ± 0.059	0.374 ± 0.067	0.430 ± 0.232	0.389 ± 0.202	0.245 ± 0.213
176	2-Nonanol	15:15.0	00:01.1	S, MS, LRI	1512	1518	0.715	0.512 ± 0.187	0.581 ± 0.344	0.429 ± 0.127	0.398 ± 0.169	0.468 ± 0.272
177	1-Nonanol	18:03.3	00:01.0	S, MS, LRI	1666	1661	0.580	1.420 ± 1.635	0.776 ± 0.538	2.515 ± 3.679	1.557 ± 2.268	1.018 ± 1.218
178	1-Heptanol	13:58.2	00:00.9	S, MS, LRI	1458	1457	0.560	1.238 ± 0.721	1.394 ± 0.657	1.120 ± 0.904	0.875 ± 0.381	1.183 ± 0.341
179	2-Methyl-1-pentanol	10:28.5	00:00.9	S, MS, LRI	1303	1297	0.229	0.262 ± 0.326	0.200 ± 0.248	0.323 ± 0.434	0.312 ± 0.145	0.199 ± 0.124
180	*trans*-4-*tert*-Butylcyclohexanol	19:43.3	00:01.1	MS, LRI	1759	1730	0.189	0.136 ± 0.326	0.149 ± 0.283	0.267 ± 0.447	0.204 ± 0.181	0.216 ± 0.332
181	2-Octanol (internal standard)	13:09.0	00:01.0	S, MS, LRI	1424	1418		40.000 ± 0.000	40.000 ± 0.000	40.000 ± 0.000	40.000 ± 0.000	40.000 ± 0.000
	Acids											
182	Propionic acid	15:43.0	00:00.7	S, MS, LRI	1536	1540	4.365	1.294 ± 0.324 [ab]	1.631 ± 0.472 [a]	1.159 ± 0.294 [b]	0.946 ± 0.158 [b]	0.991 ± 0.348 [b]
183	Acid (n.i.; *m/z* 74, 45, 73)	14:33.0	00:01.1	MS, LRI	1481	1491	4.041	0.018 ± 0.015 [b]	0.038 ± 0.020 [a]	0.016 ± 0.013 [b]	0.009 ± 0.006 [b]	0.013 ± 0.008 [b]
184	*trans*-2-Hexenoic acid	23:39.0	00:00.8	MS, LRI	1967	1967	3.651	0.081 ± 0.047 [b]	0.205 ± 0.094 [a]	0.159 ± 0.107 [ab]	0.083 ± 0.045 [b]	0.266 ± 0.213 [a]
185	Nonanoic acid	26:51.5	00:00.8	S, MS, LRI	>2100	2119	3.641	0.096 ± 0.049 [b]	0.169 ± 0.095 [b]	0.094 ± 0.043 [b]	0.166 ± 0.156 [b]	0.313 ± 0.078 [a]
186	*trans*-3-Hexenoic acid	22:49.8	00:00.8	MS, LRI	1924	1929	3.190	0.031 ± 0.033 [a]	0.006 ± 0.005 [b]	0.007 ± 0.003 [b]	0.005 ± 0.003 [b]	0.011 ± 0.005 [ab]
187	Formic acid	15:10.4	00:00.7	MS, LRI	1508	1501	2.526	1.442 ± 0.465	2.092 ± 0.849	1.523 ± 0.393	1.176 ± 0.507	1.306 ± 0.515
188	3,5,5-Trimethylhexanoic acid	23:46.0	00:00.8	MS	1973	-	2.194	0.330 ± 0.053	0.347 ± 0.115	0.370 ± 0.113	0.437 ± 0.094	0.479 ± 0.101
189	2-Propenoic acid	17:42.7	00:00.7	MS	1645	-	2.118	0.245 ± 0.073	0.262 ± 0.101	0.266 ± 0.042	0.214 ± 0.039	0.146 ± 0.055
190	Heptanoic acid	23:22.2	00:00.8	S, MS, LRI	1953	1955	1.423	0.071 ± 0.022	0.075 ± 0.044	0.083 ± 0.076	0.060 ± 0.021	0.152 ± 0.139
191	2-Decenoic acid	15:43.7	00:00.8	MS, LRI	1536	1540	1.289	0.021 ± 0.022	0.012 ± 0.024	0.005 ± 0.010	0.025 ± 0.011	0.022 ± 0.025
192	Pentanoic acid	19:34.0	00:00.8	S, MS, LRI	1751	1751	1.006	0.408 ± 0.074	0.490 ± 0.147	0.395 ± 0.074	0.394 ± 0.063	0.513 ± 0.338
193	Isobutyric acid	16:18.0	00:00.7	S, MS, LRI	1565	1555	0.832	3.347 ± 0.988	4.725 ± 1.518	4.212 ± 2.107	3.795 ± 1.757	3.200 ± 2.305
194	*trans,trans*-2,4-Hexadienoic acid	26:51.6	00:00.8	MS, LRI	>2100	2150	0.753	0.188 ± 0.145	0.053 ± 0.066	18.360 ± 48.172	4.350 ± 10.563	2.236 ± 3.744

Table 2. *Cont.*

No.	Volatile Compounds	t_R (1D) (min:s)	t_R (2D) (min:s)	ID	LRI_{exp}	LRI_{lit}	*F*-Ratio	Variety				
								MI	PO	MA	KR	SK
195	Isovaleric acid	18:18.8	00:00.7	S, MS, LRI	1683	1680	0.745	5.833 ± 1.482	3.926 ± 3.916	5.923 ± 2.842	6.130 ± 2.208	5.222 ± 2.990
196	2-Ethylhexanoic acid	23:18.0	00:00.8	MS, LRI	1949	1960	0.568	0.602 ± 1.256	0.232 ± 0.224	0.312 ± 0.271	0.144 ± 0.061	0.151 ± 0.019
197	Butyric acid	17:28.0	00:00.7	S, MS, LRI	1629	1626	0.546	18.241 ± 2.608	19.305 ± 5.118	17.622 ± 2.372	16.638 ± 0.905	18.144 ± 6.443
	Esters											
198	Methyl octanoate	12:34.0	00:01.5	MS, LRI	1401	1404	12.568	51.242 ± 14.675 ^{a}	14.588 ± 8.497 ^{b}	25.349 ± 13.750 ^{b}	21.407 ± 4.375 ^{b}	12.261 ± 13.962 ^{b}
199	*cis*-3-Hexen-1-yl acetate	10:56.0	00:01.3	MS, LRI	1324	1300	12.068	20.041 ± 8.968 ^{b}	39.339 ± 11.050 ^{a}	22.891 ± 14.489 ^{b}	6.372 ± 2.659 ^{c}	4.576 ± 3.341 ^{c}
200	Methyl hexanoate	07:51.1	00:05.4	S, MS, LRI	1183	1188	10.455	6.240 ± 2.416 ^{a}	2.417 ± 1.319 ^{b}	3.001 ± 1.098 ^{b}	2.216 ± 0.469 ^{b}	1.222 ± 1.089 ^{b}
201	Butyl hexanoate	13:05.2	00:01.7	S, MS, LRI	1422	1428	10.423	0.059 ± 0.026 ^{a}	0.015 ± 0.014 ^{b}	0.017 ± 0.016 ^{b}	0.015 ± 0.008 ^{b}	0.008 ± 0.007 ^{b}
202	Isoamyl hexanoate	14:05.0	00:01.8	S, MS, LRI	1462	1458	9.888	7.130 ± 1.573 ^{a}	1.804 ± 1.100 ^{c}	3.941 ± 3.105 ^{b}	3.323 ± 0.756 ^{bc}	1.537 ± 1.597 ^{c}
203	Ethyl 3-nonenoate	16:44.2	00:01.6	MS	1587	-	7.481	0.067 ± 0.040 ^{a}	0.021 ± 0.020 ^{b}	0.011 ± 0.005 ^{b}	0.017 ± 0.015 ^{b}	0.006 ± 0.007 ^{b}
204	Hexanodibutyrin	12:20.0	00:01.7	MS	1390	-	7.472	0.444 ± 0.181 ^{a}	0.571 ± 0.245 ^{a}	0.180 ± 0.173 ^{b}	0.179 ± 0.103 ^{b}	0.705 ± 0.336 ^{a}
205	Methyl decanoate	16:53.0	00:01.6	MS, LRI	1594	1593	7.131	1.375 ± 0.608 ^{a}	0.552 ± 0.226 ^{b}	0.741 ± 0.248 ^{b}	0.599 ± 0.132 ^{b}	0.426 ± 0.345 ^{b}
206	Phenethyl formate	20:29.8	00:01.1	MS, LRI	1799	1806	6.718	0.088 ± 0.035 ^{c}	0.211 ± 0.070 ^{a}	0.156 ± 0.046 ^{b}	0.119 ± 0.043 ^{bc}	0.112 ± 0.032 ^{bc}
207	Ethyl *trans*-4-octenoate	15:15.0	00:01.5	MS	1512	-	6.256	0.031 ± 0.003 ^{a}	0.017 ± 0.009 ^{b}	0.023 ± 0.006 ^{b}	0.016 ± 0.007 ^{b}	0.017 ± 0.006 ^{b}
208	Ethyl methyl succinate	17:36.4	00:01.1	MS, LRI	1638	1631	6.026	0.347 ± 0.156 ^{a}	0.306 ± 0.126 ^{a}	0.242 ± 0.078 ^{a}	0.079 ± 0.067 ^{b}	0.254 ± 0.070 ^{a}
209	Octyl formate	16:04.2	00:01.0	MS, LRI	1553	1560	5.909	6.604 ± 0.947 ^{b}	6.267 ± 1.160 ^{b}	6.531 ± 1.465 ^{b}	9.177 ± 1.345 ^{a}	6.050 ± 2.280 ^{b}
210	*trans*-3-Hexen-1-yl acetate	10:42.0	00:01.3	MS, LRI	1313	1316	5.675	20.322 ± 7.276 ^{a}	10.073 ± 8.187 ^{b}	5.922 ± 3.176 ^{b}	10.530 ± 4.255 ^{b}	7.388 ± 8.444 ^{b}
211	Ethyl hexadecanoate	30:13.7	00:01.5	MS, LRI	>2100	2261	5.586	0.560 ± 0.414 ^{a}	0.051 ± 0.082 ^{b}	0.119 ± 0.222 ^{b}	0.091 ± 0.149 ^{b}	0.052 ± 0.064 ^{b}
212	Ethyl 2-hydroxy-4-methylpentanoate	15:49.3	00:01.0	MS	1541	-	5.543	0.380 ± 0.161 ^{c}	0.829 ± 0.571 ^{bc}	1.875 ± 0.999 ^{a}	1.625 ± 1.112 ^{ab}	0.266 ± 0.231 ^{c}
213	2-Phenylethyl isobutyrate	22:45.7	00:01.1	MS, LRI	1921	1916	5.419	3.788 ± 1.880 ^{a}	1.824 ± 1.708 ^{b}	1.421 ± 1.243 ^{b}	0.816 ± 0.397 ^{b}	0.769 ± 0.245 ^{b}
214	Propyl hexanoate	10:56.2	00:01.7	MS, LRI	1324	1319	4.982	0.760 ± 0.330 ^{a}	0.339 ± 0.286 ^{b}	0.286 ± 0.227 ^{b}	0.379 ± 0.143 ^{b}	0.170 ± 0.180 ^{b}
215	Diethyl glutarate	20:23.0	00:01.2	MS, LRI	1793	1780	4.706	0.028 ± 0.012 ^{c}	0.105 ± 0.061 ^{a}	0.064 ± 0.043 ^{bc}	0.059 ± 0.013 ^{bc}	0.092 ± 0.020 ^{ab}
216	3-Ethoxypropyl acetate	11:52.0	00:01.2	MS	1368	-	4.423	0.754 ± 0.514 ^{b}	2.176 ± 1.442 ^{a}	0.638 ± 0.941 ^{b}	0.442 ± 0.365 ^{b}	0.454 ± 0.707 ^{b}
217	Isoamyl octanoate	18:06.1	00:01.9	MS, LRI	1669	1657	4.033	4.133 ± 0.747 ^{a}	2.340 ± 0.822 ^{b}	3.155 ± 1.362 ^{ab}	3.789 ± 0.950 ^{a}	2.125 ± 1.656 ^{b}
218	Ethyl heptanoate	11:17.0	00:01.6	MS, LRI	1341	1342	3.762	3.007 ± 1.131 ^{a}	1.591 ± 1.355 ^{b}	1.701 ± 0.709 ^{b}	1.498 ± 0.355 ^{b}	0.938 ± 1.126 ^{b}
219	Ethyl pyruvate	09:53.4	00:01.0	MS, LRI	1276	1276	3.717	1.711 ± 0.693 ^{b}	3.025 ± 0.908 ^{a}	1.931 ± 1.095 ^{b}	1.496 ± 0.650 ^{b}	1.685 ± 0.421 ^{b}
220	Isoamyl decanoate	21:58.0	00:01.9	MS, LRI	1879	1871	3.685	0.039 ± 0.093 ^{b}	0.191 ± 0.114 ^{ab}	0.180 ± 0.153 ^{ab}	0.239 ± 0.042 ^{ab}	0.178 ± 0.115 ^{ab}
221	Propyl octanoate	15:22.0	00:01.8	MS, LRI	1518	1504	3.637	0.893 ± 0.277 ^{a}	0.428 ± 0.431 ^{b}	0.384 ± 0.269 ^{b}	0.577 ± 0.297 ^{ab}	0.270 ± 0.261 ^{b}
222	Ethyl 4-pyrazolecarboxylate	14:33.0	00:01.1	MS	1481	-	3.291	0.016 ± 0.002 ^{b}	0.026 ± 0.011 ^{a}	0.014 ± 0.010 ^{b}	0.012 ± 0.008 ^{b}	0.010 ± 0.009 ^{b}
223	Methyl 2-methyllactate	16:39.0	00:01.2	MS	1582	-	3.049	0.116 ± 0.015 ^{a}	0.088 ± 0.013 ^{bc}	0.102 ± 0.017 ^{ab}	0.109 ± 0.024 ^{ab}	0.068 ± 0.059 ^{c}
224	3-Hydroxy-2,4,4-trimethylpentyl isobutyrate	21:54.7	00:01.2	MS	1876	-	3.046	0.089 ± 0.019 ^{b}	0.191 ± 0.051 ^{a}	0.124 ± 0.108 ^{b}	0.104 ± 0.035 ^{b}	0.112 ± 0.044 ^{b}
225	Isobutyl hexanoate	11:45.0	00:01.8	MS, LRI	1362	1357	2.919	1.039 ± 0.567 ^{a}	0.392 ± 0.297 ^{b}	0.767 ± 0.580 ^{ab}	0.595 ± 0.202 ^{ab}	0.259 ± 0.272 ^{b}
226	Hydroxyl acid ester (n.i.; *m/z* 143, 115, 75)	17:49.0	00:01.3	MS	1651	-	2.680	0.005 ± 0.005	0.012 ± 0.005	0.009 ± 0.006	0.007 ± 0.004	0.014 ± 0.005
227	Ethyl 4-hydroxybutyrate	20:47.7	00:00.9	MS, LRI	1815	1819	2.518	0.955 ± 0.717	1.134 ± 0.677	1.141 ± 0.695	0.315 ± 0.171	0.536 ± 0.197
228	Octyl acetate	14:24.4	00:01.5	MS, LRI	1475	1475	2.454	0.408 ± 0.857	1.601 ± 1.141	1.423 ± 1.416	3.153 ± 1.639	6.579 ± 10.545

Table 2. *Cont.*

No.	Volatile Compounds	t_R (1D) (min:s)	t_R (2D) (min:s)	ID	LRI_{exp}	LRI_{lit}	*F*-Ratio	Variety				
								MI	PO	MA	KR	SK
229	3-Methyl-3-buten-1-yl acetate	08:07.8	00:01.2	MS, LRI	1196	1190	2.448	0.072 ± 0.037	0.066 ± 0.036	0.032 ± 0.028	0.049 ± 0.021	0.025 ± 0.026
230	Ethyl 2-octenoate	16:04.0	00:01.5	MS, LRI	1553	1557	2.386	0.031 ± 0.005	0.027 ± 0.009	0.024 ± 0.008	0.019 ± 0.003	0.027 ± 0.015
231	Ethyl undecanoate	18:45.2	00:01.0	MS, LRI	1709	1725	2.352	0.539 ± 0.749	0.198 ± 0.111	0.319 ± 0.391	1.006 ± 0.777	0.166 ± 0.169
232	Pentyl acetate	07:36.1	00:01.3	S, MS, LRI	1172	1161	2.334	0.123 ± 0.199	0.088 ± 0.204	0.037 ± 0.056	0.282 ± 0.174	0.053 ± 0.057
233	Isobutyl octanoate	15:57.7	00:01.9	MS, LRI	1548	1551	2.226	0.217 ± 0.071	0.119 ± 0.083	0.185 ± 0.114	0.213 ± 0.109	0.074 ± 0.069
234	Ethyl 4-hexenoate	10:21.3	00:01.4	MS, LRI	1297	1292	2.211	7.705 ± 10.236	2.044 ± 1.364	1.287 ± 0.650	0.784 ± 0.458	1.363 ± 1.374
235	Ethyl *trans*-2-butenoate	07:19.0	00:05.0	MS, LRI	1159	1161	2.127	8.551 ± 4.382	5.204 ± 3.185	7.463 ± 2.841	4.376 ± 2.710	4.293 ± 2.995
236	*trans*,*trans*-2,4-Octadien-1-yl acetate	16:18.0	00:01.4	MS	1565	-	1.956	0.017 ± 0.017	0.024 ± 0.018	0.006 ± 0.010	0.007 ± 0.009	0.017 ± 0.017
237	2-Phenylethyl octanoate	32:17.5	00:01.3	MS, LRI	>2100	2373	1.762	0.032 ± 0.036	0.006 ± 0.005	0.009 ± 0.004	0.024 ± 0.030	0.004 ± 0.004
238	Isoamyl butyrate	09:45.8	00:01.7	S, MS, LRI	1270	1266	1.756	2.573 ± 0.919	2.131 ± 0.957	2.055 ± 0.892	1.912 ± 0.358	1.089 ± 1.161
239	Di-isobutyl acetate	20:02.0	00:01.2	MS	1775	-	1.724	0.206 ± 0.065	0.243 ± 0.211	0.139 ± 0.106	0.112 ± 0.037	0.094 ± 0.032
240	3-Methylheptyl acetate	12:26.5	00:01.6	MS	1395	-	1.665	0.367 ± 0.347	0.236 ± 0.216	0.182 ± 0.086	0.112 ± 0.050	0.124 ± 0.077
241	Diethyl malonate	16:32.0	00:01.1	MS, LRI	1577	1574	1.648	0.174 ± 0.052	0.188 ± 0.079	0.148 ± 0.050	0.126 ± 0.033	0.197 ± 0.034
242	Heptyl acetate	12:13.4	00:01.5	MS, LRI	1385	1385	1.487	0.236 ± 0.133	0.249 ± 0.179	0.151 ± 0.091	0.120 ± 0.044	0.143 ± 0.145
243	Ethyl hydrogen succinate	30:16.2	00:00.8	MS, LRI	>2100	2350	1.443	3.279 ± 0.920	7.386 ± 6.573	5.875 ± 1.785	4.985 ± 2.470	4.242 ± 1.226
244	Methyl 2-isopropoxypropanoate	20:30.0	00:01.3	MS	1799	-	1.288	0.030 ± 0.042	0.076 ± 0.036	0.061 ± 0.043	0.065 ± 0.045	0.053 ± 0.047
245	Vinyl decanoate	19:27.2	00:01.4	MS	1745	-	1.258	0.224 ± 0.357	0.086 ± 0.065	0.044 ± 0.038	0.038 ± 0.034	0.130 ± 0.062
246	Diethyl malate	25:29.4	00:00.9	MS, LRI	2063	2065	1.210	0.227 ± 0.144	0.462 ± 0.502	0.399 ± 0.220	0.360 ± 0.231	0.629 ± 0.235
247	*trans*-Penten-1-yl acetate	08:50.0	00:01.2	MS	1228	-	1.132	0.050 ± 0.049	0.039 ± 0.049	0.024 ± 0.031	0.011 ± 0.011	0.024 ± 0.023
248	Phenylmethyl acetate	19:27.0	00:01.2	MS, LRI	1745	1747	1.119	0.042 ± 0.026	0.211 ± 0.394	0.035 ± 0.018	0.046 ± 0.018	0.054 ± 0.021
249	Ethyl 3-hydroxybutyrate	15:15.5	00:00.9	MS, LRI	1512	1512	1.045	0.398 ± 0.278	0.407 ± 0.202	0.430 ± 0.195	0.233 ± 0.157	0.430 ± 0.093
250	Isoamyl propanoate	07:56.8	00:01.5	MS, LRI	1188	1188	1.017	0.638 ± 0.264	0.677 ± 0.436	0.740 ± 0.490	0.543 ± 0.341	0.250 ± 0.186
251	Ethyl hydroxyacetate	13:09.0	00:00.8	MS, LRI	1424	1436	0.987	0.032 ± 0.045	0.097 ± 0.079	0.058 ± 0.076	0.085 ± 0.144	0.163 ± 0.215
252	Ethyl nonanoate	15:43.0	00:01.7	MS, LRI	1536	1535	0.928	1.843 ± 0.503	0.984 ± 0.394	0.971 ± 0.345	2.321 ± 3.355	1.116 ± 0.807
253	2-(1,1-Dimethylethyl) -cyclohexen-1-yl acetate	16:11.0	00:01.8	MS	1559	-	0.848	0.046 ± 0.026	0.024 ± 0.015	0.045 ± 0.031	0.040 ± 0.032	0.034 ± 0.004
254	Ethyl 2-hydroxy-3-phenylpropanoate	29:11.1	00:01.0	MS, LRI	>2100	2273	0.846	0.001 ± 0.001	0.001 ± 0.001	0.009 ± 0.019	0.007 ± 0.014	0.001 ± 0.000
255	Ethyl 3-methylbutylbutanedioate	22:29.7	00:01.3	MS, LRI	1907	1907	0.835	2.315 ± 0.806	3.353 ± 2.324	2.743 ± 1.814	2.843 ± 0.908	1.639 ± 0.545
256	Ethyl 9-decenoate	18:45.0	00:01.6	S, MS, LRI	1708	1708	0.801	0.133 ± 0.189	0.177 ± 0.227	0.280 ± 0.347	0.101 ± 0.086	0.067 ± 0.058
257	Ethyl 3-ethoxy-*trans*-2-propenoate	15:43.0	00:01.2	MS	1536	-	0.785	1.418 ± 0.096	1.363 ± 0.205	1.480 ± 0.136	1.542 ± 0.317	1.360 ± 0.353
258	Butyl ethyl succinate	20:37.0	00:01.3	MS, LRI	1806	1820	0.776	0.245 ± 0.112	0.295 ± 0.195	0.294 ± 0.182	0.220 ± 0.139	0.136 ± 0.063
259	Ethyl 2,4-hexadienoate I	14:35.0	00:01.3	MS, LRI	1483	1501	0.673	0.033 ± 0.034	0.019 ± 0.011	0.869 ± 2.269	0.343 ± 0.855	0.097 ± 0.153
260	Ethyl 2,4-hexadienoate II	15:08.0	00:01.3	MS, LRI	1506	1501	0.645	0.290 ± 0.328	0.038 ± 0.016	6.200 ± 16.289	3.473 ± 8.685	0.353 ± 0.515
261	Isobutyl acetate	04:23.8	00:04.7	S, MS, LRI	1015	1009	0.531	0.726 ± 0.371	0.774 ± 0.280	0.728 ± 0.435	0.534 ± 0.175	0.658 ± 0.405
262	2-Phenylethyl isovalerate	24:00.0	00:01.4	MS, LRI	1985	1988	0.530	0.015 ± 0.012	0.014 ± 0.014	0.013 ± 0.010	0.020 ± 0.006	0.011 ± 0.010
263	Methyl 2-hydroxybutanoate	11:45.0	00:01.1	MS, LRI	1362	1382	0.529	0.274 ± 0.056	0.217 ± 0.155	0.208 ± 0.156	0.280 ± 0.040	0.225 ± 0.197

Table 2. *Cont.*

No.	Volatile Compounds	t_R (1D)	t_R (2D)	ID	LRI_{exp}	LRI_{lit}	F-Ratio	Variety				
		(min:s)	(min:s)					MI	PO	MA	KR	SK
264	Isopropyl lactate	15:36.0	00:01.6	MS	1530	-	0.512	0.032 ± 0.007	0.027 ± 0.004	0.029 ± 0.016	0.026 ± 0.007	0.023 ± 0.024
265	Ethyl *cis*-4-decenoate	18:17.4	00:01.6	MS, LRI	1681	1680	0.510	0.017 ± 0.019	0.012 ± 0.010	0.008 ± 0.014	0.015 ± 0.011	0.011 ± 0.009
266	Ethyl *cis*-4-octenoate	14:39.6	00:01.5	MS	1486	-	0.503	0.127 ± 0.073	0.094 ± 0.054	0.123 ± 0.053	0.098 ± 0.015	0.101 ± 0.094
267	Ethyl 2-hexenoate	11:31.7	00:01.5	S, MS, LRI	1352	1357	0.469	1.626 ± 0.984	1.864 ± 0.759	2.107 ± 1.213	1.408 ± 0.806	1.721 ± 1.542
268	Isoamyl lactate	16:18.0	00:01.0	MS, LRI	1565	1572	0.376	0.449 ± 0.304	0.829 ± 1.062	0.597 ± 0.380	0.703 ± 0.581	0.596 ± 0.341
269	Ethyl 2-propynoate	09:11.0	00:01.8	MS	1244	-	0.368	4.550 ± 1.065	4.324 ± 1.444	4.768 ± 1.504	4.322 ± 1.565	3.562 ± 2.475
270	Diethyl fumarate	17:56.0	00:01.2	MS, LRI	1659	1660	0.155	0.055 ± 0.027	0.047 ± 0.018	0.048 ± 0.031	0.050 ± 0.014	0.045 ± 0.014
	Volatile phenols											
271	2-Methoxyphenol	21:47.0	00:00.9	MS, LRI	1869	1869	5.084	0.023 ± 0.016 [b]	0.052 ± 0.034 [a]	0.019 ± 0.011 [b]	0.010 ± 0.001 [b]	0.011 ± 0.003 [b]
272	4-Vinylguaiacol	27:33.5	00:00.9	S, MS, LRI	>2100	2168	2.970	0.563 ± 0.266 [ab]	0.762 ± 0.426 [a]	0.373 ± 0.170 [b]	0.377 ± 0.330 [b]	0.163 ± 0.069 [b]
273	Phenol	24:14.0	00:00.8	S, MS, LRI	1998	1995	1.607	0.701 ± 0.049	0.856 ± 0.131	0.804 ± 0.309	0.601 ± 0.130	0.886 ± 0.541
274	4-Ethylguaiacol	24:44.7	00:01.0	S, MS, LRI	2024	2024	1.514	0.011 ± 0.020	0.024 ± 0.034	0.004 ± 0.004	0.002 ± 0.005	0.003 ± 0.003
275	2,4-Bis(1,1-dimethylethyl) phenol	29:01.0	00:01.0	MS, LRI	>2100	2270	1.328	0.923 ± 0.417	0.821 ± 0.247	1.133 ± 0.583	0.667 ± 0.280	0.810 ± 0.189
	Furanoids and lactones											
276	3-Methyl-2(5H)-furanone	19:13.3	00:01.0	MS, LRI	1733	1726	7.776	0.029 ± 0.012 [a]	0.024 ± 0.011 [ab]	0.017 ± 0.009 [b]	0.004 ± 0.004 [c]	0.017 ± 0.006 [ab]
277	Acetylfuran	15:01.2	00:01.0	MS, LRI	1501	1501	4.958	0.092 ± 0.054 [b]	0.416 ± 0.326 [a]	0.201 ± 0.104 [b]	0.086 ± 0.040 [b]	0.092 ± 0.031 [b]
278	γ-Butyrolactone	17:28.0	00:01.0	MS, LRI	1629	1626	4.491	3.839 ± 1.256 [a]	4.284 ± 1.527 [a]	4.055 ± 0.852 [a]	1.918 ± 1.271 [b]	2.661 ± 0.844 [ab]
279	2-(5-Methyl-5-vinyltetra hydro-2-furanyl) -2-propanol	14:16.1	00:01.2	MS	1470	-	4.293	0.050 ± 0.089 [b]	0.106 ± 0.063 [b]	0.053 ± 0.063 [b]	0.100 ± 0.084 [b]	0.305 ± 0.233 [a]
280	γ-Nonalactone	24:42.6	00:01.2	S, MS, LRI	2022	2018	3.909	0.037 ± 0.072 [c]	0.311 ± 0.222 [a]	0.152 ± 0.123 [bc]	0.200 ± 0.100 [ab]	0.191 ± 0.089 [abc]
281	Lactone (n.i.; *m/z* 85, 57, 100)	23:39.0	00:01.1	MS	1967	-	2.450	0.057 ± 0.059	0.013 ± 0.014	0.039 ± 0.037	0.010 ± 0.006	0.009 ± 0.001
282	Pantolactone	24:42.0	00:00.8	MS, LRI	2022	2029	2.220	0.091 ± 0.018	0.184 ± 0.075	0.174 ± 0.101	0.156 ± 0.067	0.123 ± 0.046
283	Furfuryl ether	10:14.0	00:01.1	MS	1292	-	2.138	0.159 ± 0.187	0.242 ± 0.117	0.245 ± 0.100	0.091 ± 0.033	0.116 ± 0.071
284	Furfural	14:05.2	00:00.9	S, MS, LRI	1462	1460	2.087	1.122 ± 0.381	16.951 ± 26.840	2.258 ± 1.044	0.857 ± 0.168	1.010 ± 0.306
285	Ethyl 2-furoate	17:21.0	00:01.1	MS, LRI	1622	1624	1.823	4.805 ± 1.288	7.089 ± 2.513	4.929 ± 2.119	5.484 ± 1.375	4.786 ± 1.742
286	γ-Octalactone	22:49.8	00:01.1	S, MS, LRI	1924	1923	1.722	0.533 ± 0.208	0.907 ± 0.309	0.708 ± 0.325	0.723 ± 0.299	0.730 ± 0.048
287	2(5H)-furanone	19:59.4	00:00.9	S, MS, LRI	1773	1787	1.266	0.079 ± 0.013	0.151 ± 0.172	0.078 ± 0.013	0.060 ± 0.010	0.079 ± 0.046
288	5-Methyl-2-furfural	16:25.8	00:01.0	S, MS, LRI	1571	1570	1.263	0.014 ± 0.014	1.971 ± 4.278	0.055 ± 0.053	0.033 ± 0.013	0.027 ± 0.033
289	Lactone (n.i.; *m/z* 99, 71, 87)	23:41.8	00:01.2	MS	1970	-	1.206	0.012 ± 0.022	0.121 ± 0.223	0.022 ± 0.036	0.025 ± 0.022	0.040 ± 0.034
290	γ-Hydroxymethyl-γ-butyrolactone	28:33.0	00:00.9	MS	>2100	-	0.725	1.594 ± 1.045	1.857 ± 1.170	2.753 ± 1.813	2.442 ± 1.912	1.735 ± 1.520
291	5-Ethoxydihydro-2(3H)-furanone	19:20.0	00:01.0	MS, LRI	1739	1728	0.666	0.054 ± 0.024	0.059 ± 0.031	0.071 ± 0.032	0.046 ± 0.027	0.062 ± 0.039
292	δ-Caprolactone	20:37.4	00:01.1	MS, LRI	1806	1818	0.121	0.344 ± 0.225	0.332 ± 0.181	0.357 ± 0.160	0.305 ± 0.174	0.288 ± 0.108
	Sulfur containing compounds											
293	Methional	14:02.7	00:01.0	MS, LRI	1461	1461	11.821	0.018 ± 0.014 [c]	0.116 ± 0.047 [a]	0.060 ± 0.043 [b]	0.017 ± 0.005 [c]	0.027 ± 0.031 [bc]
294	2-(Methylthio)ethanol	15:29.0	00:00.8	S, MS, LRI	1524	1531	9.501	0.261 ± 0.063 [b]	0.356 ± 0.069 [a]	0.290 ± 0.094 [ab]	0.133 ± 0.020 [c]	0.237 ± 0.103 [b]
295	Methionol	19:08.8	00:00.9	S, MS, LRI	1729	1733	5.647	2.344 ± 0.660 [bc]	4.022 ± 1.550 [a]	3.056 ± 1.076 [ab]	1.741 ± 0.881 [c]	1.465 ± 0.676 [c]

Table 2. *Cont.*

No.	Volatile Compounds	t_R (1D) (min:s)	t_R (2D) (min:s)	ID	LRI_{exp}	LRI_{lit}	*F*-Ratio	Variety MI	PO	MA	KR	SK
296	Ethyl thiophene-2-carboxylate	20:02.0	00:01.2	MS	1775	-	4.883	0.024 ± 0.006^{a}	0.020 ± 0.003^{ab}	0.018 ± 0.005^{bc}	0.017 ± 0.004^{bc}	0.012 ± 0.002^{c}
297	4-(Methylthio)-1-butanol	21:26.0	00:00.9	MS	1850	-	4.672	0.011 ± 0.005^{bc}	0.022 ± 0.010^{a}	0.018 ± 0.009^{ab}	0.009 ± 0.004^{c}	0.007 ± 0.003^{c}
298	S-(3-hydroxypropyl) thioacetate	14:47.0	00:01.1	MS	1491	-	4.320	0.052 ± 0.015^{b}	0.083 ± 0.032^{a}	0.062 ± 0.016^{b}	0.048 ± 0.006^{b}	0.042 ± 0.020^{b}
299	2-Thiophenecarboxaldehyde	18:45.0	00:01.0	S, MS, LRI	1708	1701	3.796	0.039 ± 0.017^{b}	0.087 ± 0.046^{a}	0.063 ± 0.036^{ab}	0.032 ± 0.013^{b}	0.069 ± 0.020^{ab}
300	Ethyl methanesulfonate	18:31.0	00:00.9	MS	1695	-	3.638	0.177 ± 0.028^{a}	0.157 ± 0.037^{a}	0.159 ± 0.030^{a}	0.120 ± 0.023^{b}	0.163 ± 0.029^{a}
301	Ethyl 3-(methylthio)-*trans*-2-propenoate	19:41.0	00:01.2	MS, LRI	1757	1733	2.915	0.013 ± 0.007^{b}	0.021 ± 0.008^{a}	0.012 ± 0.004^{b}	0.018 ± 0.003^{ab}	0.010 ± 0.007^{b}
302	3-(Methylthio)propyl acetate	17:28.7	00:01.2	MS, LRI	1630	1627	2.832	0.589 ± 0.319^{ab}	0.731 ± 0.337^{a}	0.385 ± 0.305^{bc}	0.427 ± 0.061^{bc}	0.202 ± 0.087^{c}
303	Ethyl 3-(methylthio)-*trans*-2-propenoate	21:33.0	00:01.2	MS, LRI	1856	1837	2.795	0.002 ± 0.004^{b}	0.011 ± 0.006^{a}	0.006 ± 0.005^{ab}	0.010 ± 0.006^{a}	0.005 ± 0.006^{ab}
304	Diethyl sulfate	17:21.0	00:01.0	MS	1622	-	2.496	0.013 ± 0.003	0.010 ± 0.004	0.011 ± 0.003	0.009 ± 0.002	0.011 ± 0.002
305	Ethyl thiocyanate	21:05.2	00:01.2	MS	1831	-	2.399	0.250 ± 0.090	0.299 ± 0.119	0.279 ± 0.086	0.251 ± 0.073	0.112 ± 0.045
306	3-Ethoxythiophene	14:19.0	00:01.2	MS	1472	-	1.656	0.030 ± 0.015	0.042 ± 0.035	0.054 ± 0.039	0.023 ± 0.005	0.020 ± 0.018
307	S-[(2,5-dihydro-4-hydroxy-5 -oxo-3-furanyl)methyl] ethanethioate	23:09.7	00:00.8	MS	1942	-	1.376	0.113 ± 0.149	0.032 ± 0.055	0.048 ± 0.057	0.030 ± 0.024	0.016 ± 0.020
308	1-(*tert*-Butylsulfonyl)-2 -octanol	19:14.9	00:02.2	MS	1734	-	0.996	0.182 ± 0.125	0.193 ± 0.116	0.220 ± 0.078	0.160 ± 0.086	0.086 ± 0.078
309	Cyclohexyl isothiocyanate	18:17.0	00:01.6	MS, LRI	1681	1667	0.780	0.014 ± 0.011	0.017 ± 0.011	0.017 ± 0.009	0.014 ± 0.008	0.006 ± 0.005
310	3-[(2-Hydroxyethyl)thio]-1 -propanol	20:58.0	00:00.9	MS	1825	-	0.515	0.020 ± 0.004	0.019 ± 0.006	0.027 ± 0.023	0.029 ± 0.026	0.022 ± 0.009
311	2-Methyldihydro-3(2H)-thiophenone	15:28.6	00:01.1	MS, LRI	1523	1538	0.510	1.630 ± 0.556	1.418 ± 0.673	1.130± 0.911	1.449 ± 0.752	1.235 ± 0.462
	Other compounds											
312	2,6,10,10-Tetramethyl-1-oxaspiro[4.5]deca-3,6-diene	15:50.0	00:01.9	MS	1541	-	7.995	0.139 ± 0.058^{a}	0.025 ± 0.012^{c}	0.076 ± 0.036^{b}	0.054 ± 0.020^{bc}	0.075 ± 0.072^{bc}
313	Ethylene diglycol monoethyl ether	17:19.0	00:00.9	MS, LRI	1620	1622	5.688	0.081 ± 0.031^{b}	0.238 ± 0.114^{a}	0.175 ± 0.095^{a}	0.227 ± 0.062^{a}	0.261 ± 0.017^{a}
314	Acetic formic anhydride	15:59.6	00:00.7	MS	1549	-	1.865	0.065 ± 0.090	0.159 ± 0.132	0.151 ± 0.102	0.059 ± 0.084	0.040 ± 0.053
315	Crotonic anhydride	21:25.8	00:01.5	MS	1850	-	1.557	0.077 ± 0.060	0.063 ± 0.042	0.043 ± 0.023	0.065 ± 0.018	0.020 ± 0.017
316	1H-indole	31:00.2	00:00.9	MS, LRI	>2100	2420	1.036	0.044 ± 0.012	0.041 ± 0.019	0.037 ± 0.021	0.027 ± 0.005	0.035 ± 0.026
317	Methylsuccinic anhydride	18:59.0	00:00.9	MS	1720	-	0.517	0.013 ± 0.021	0.051 ± 0.112	0.038 ± 0.028	0.029 ± 0.026	0.053 ± 0.009

ID—identification of compounds; S—retention time and mass spectrum consistent with that of the pure standard and with NIST05 mass spectra electronic library; LRI—linear retention index consistent with that found in literature; MS—mass spectra consistent with that from NIST 2.0, Wiley 8, and FFNSC 2 mass spectra electronic libraries or literature; n.i.—not identified. The compounds with only MS symbol in ID column were tentatively identified. LRI_{lit}—linear retention index from the literature, LRI_{exp}—linear retention index obtained experimentally. Varieties: MI—Malvazija istarska, PO—Pošip, MA—Maraština, KR—Kraljevina, SK—Škrlet. Different superscript lowercase letters in a row represent statistically significant differences between mean values at $p < 0.05$ obtained by one-way ANOVA and least significant difference (LSD) test.

No significant differences were found between the concentrations of hydrocarbons, while aldehydes also turned out to be poor varietal differentiators, with significant differences found only for decanal (Table 2). On the other hand, several ketones were found useful for this purpose: the highest concentration of 2-undecanone and 3-undecanone was specific for Malvazija istarska, 1,4,7,10,13-pentaoxacyclononadecane-14,19-dione and cyclohexylideneacetone were characteristic for Škrlet, while the lowest concentration of isophorone was found in Maraština wines.

4-Methyl-1-heptanol was the most useful among alcohols in differentiating monovarietal wines with a rather high *F*-ratio (Table 2). It was found in the highest concentration in Škrlet, followed by Malvazija istarska wines, while the other wines contained lower concentrations. The results for *cis*-3-hexen-1-ol were in accordance with those obtained by GC-MS, with the highest concentration found in Dalmatian Pošip and Maraština wines. 3-Octanol and 1-octen-3-ol were exclusive markers for Pošip, 2-decanol for Škrlet, while the lowest concentration of an isomer of 2-penten-1-ol was characteristic for Kraljevina wine. *F*-ratios determined for fatty acids were relatively low and significant differences were found only for five of them.

A very large number of minor esters was identified by GC×GC-TOF-MS analysis (Table 2). In accordance with the GC-MS data, the concentrations of the majority of esters of aliphatic higher alcohols and fatty acids were the highest in Malvazija istarska wines. Despite the thesis that precursor concentrations do not significantly determine the concentrations of acetate esters formed by *Saccharomyces cerevisiae* [60], the highest concentration of *cis*-3-hexen-1-yl acetate corresponded to the highest concentration of its precursor, *cis*-3-hexen-1-ol, found in Pošip wine. Pošip wine was the most abundant in particular esters of ethanol and hydroxyl keto acids, such as diethyl glutarate and ethyl pyruvate. Although without a statistically significant difference, the concentrations of the related esters, such as ethyl lactate and diethyl succinate, determined by GC-MS, also had a tendency to be higher in Pošip wines.

Pošip wines contained the highest concentration of volatile phenols, such as 2-methoxyphenol and 4-vinylguaiacol. Significant differences were found for particular furanoids and lactones. A number of sulfur containing compounds was identified, with many of them found in the highest concentration in Pošip wines, some with relatively high *F*-ratios, such as methional. Kraljevina and Škrlet wines were generally the least abundant in these compounds (Table 2).

3.3. Multivariate Statistical Analysis

PCA allowed a good separation of the investigated monovarietal wines according to variety when applied on a dataset reduced to 40 variables with the highest *F*-values, obtained by both GC-MS and GC×GC-TOF-MS analysis. Monovarietal wines were clearly separated from each other in two-dimensional space despite a relatively high number of varieties (Figure 1). Škrlet wine was clearly differentiated from the others along the direction of PC1 and was characterized by higher amounts of terpenes. A part of Malvazija istarska wines also gravitated towards higher positive PC1 values, but the wines of this variety were also separated from the others along the direction of PC2, mostly due to higher concentrations of particular esters with positive PC2 values. Volatile aroma compounds located in the second quadrant of Cartesian system with negative PC1 and positive PC2 coordinates, 2,3-dihydro-1,1,5,6-tetramethyl-1H-indene and γ-dehydro-ar-himachalene, contributed most to the separation of Kraljevina wines, while the location of Pošip wines was obviously conditioned by the loadings of *cis*-3-hexen-1-ol, vitispirane II, ethyl benzoate, methional, *cis*-3-hexen-1-yl acetate, 2-phenethyl acetate, and 2-(methylthio)ethanol. Maraština wines were apparently not linked to any particular compound class, probably due to lower concentrations of the 40 volatile compounds used for PCA.

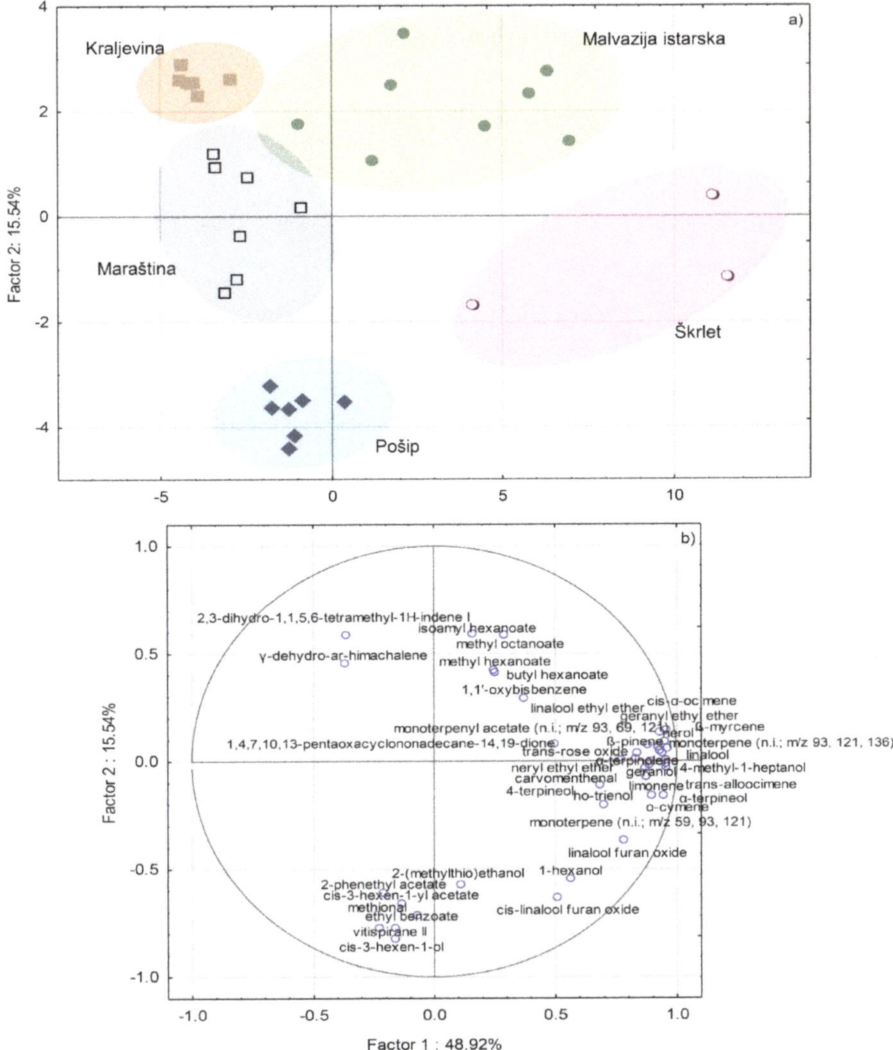

Figure 1. (**a**) Separation of Croatian monovarietal wines according to variety in two-dimensional space defined by the first two principal components, PC1 and PC2; (**b**) Factor loadings of selected variables (40 volatile aroma compounds with the highest *F*-ratios), as determined by gas chromatography mass spectrometry (GC-MS) and two-dimensional gas chromatography with time-of-flight mass spectrometry (GC×GC-TOF-MS) analysis, on PC1 and PC2.

Hierarchical clustering analysis according to variety, performed using the amounts of the 60 volatile aroma compounds with the highest *F*-ratio, confirmed that each monovarietal wine had a distinct volatile profile (Figure 2). Most of the conclusions were similar to those obtained by the PCA. Škrlet and Malvazija Istarska wines were clearly separated from each other mostly due to higher concentrations of particular esters in the latter, but were clustered together by high terpene concentrations. The generated heatmap probably offered the clearest insight into the intra-varietal diversity of particular wines, especially Malvazija with two evident clusters with different terpene content. Pošip formed a distinct cluster mostly due to high concentrations of particular compounds from several classes, some of them already mentioned in the PCA, including vitispirane II, *trans*-edulan, methional, 2-phenyletahnol, *cis*-3-hexen-1-ol and its acetate, ethyl benzoate, 2-heptanol, 2-phenethyl acetate, ethyl cinnamate, and others. Kraljevina wines were clearly the least abundant in the majority of the 60 pre-selected compounds, except for γ-dehydro-ar-himachalene, 1,2-dihydro-1,4,6-trimethylnaphthalene and particular benzenoids, which were confirmed as its markers.

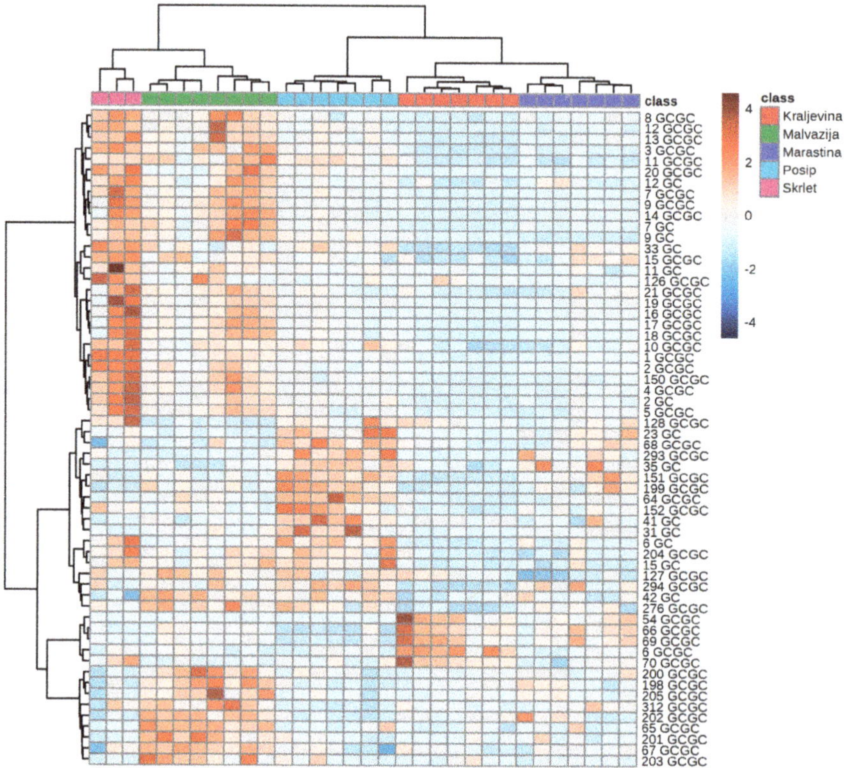

Figure 2. Hierarchical clustering analysis performed using volatile aroma compound profiles of Croatian monovarietal wines obtained by GC-MS and GC×GC-TOF-MS analysis. The heatmap was generated using 60 most significant compounds (the highest *F*-ratios). The rows in the heatmap represent compounds and the columns indicate samples. Compounds are designated by numbers which correspond to those in Table 1 (GC, i.e., GC-MS) or in Table 2 (GCGC, i.e., GC×GC-TOF-MS). The colors of heatmap cells indicate the abundance of compounds across different samples. The color gradient, ranging from dark blue through white to dark red, represents low, middle, and high abundance of a compound.

SLDA was applied on a dataset reduced to 60 most significant volatile aroma compounds according to *F*-ratio from both GC-MS and GC×GC-TOF-MS original datasets. All the monovarietal wines were classified correctly according to variety by this model, and 24 most significant variables were extracted (Figure 3), with rather high squared Mahalanobis distances from group centroids. A 100% correct classification was obtained after including only seven variables. α-Terpineol was confirmed once again as the most powerful varietal marker, since the SLDA model classified correctly 68.75% of all the wines and 100.00% of Škrlet wines by using only this variable. After including β-pinene and ethyl benzoate the total percentage of correctly classified wines increased to 93.75%. For achieving a 100.00% correct classification, 1,1′-oxybisbenzene, γ-dehydro-ar-himachalene, vitispirane II, and 2,6,10,10-tetramethyl-1-oxaspiro[4.5]deca-3,6-diene were included in the SLDA model. The following 17 volatile aroma compounds were also included: 2-phenethyl acetate, isophorone, monoterpenyl acetate (n.i.; *m/z* 93, 69, 121), 2,3-dihydro-1,1,5,6-tetramethyl-1H-indene II, *cis*-3-hexen-1-ol, methyl hexanoate, *trans*-rose oxide, methyl decanoate, *cis*-3-hexen-1-yl acetate, monoterpene (n.i.; *m/z* 93, 69, 41), β-myrcene, limonene, 3-methyl-2(5H)-furanone, 2-phenylethanol, 1,2-dihydro-1,4,6-trimethylnaphthalene, nerol, and nerol oxide.

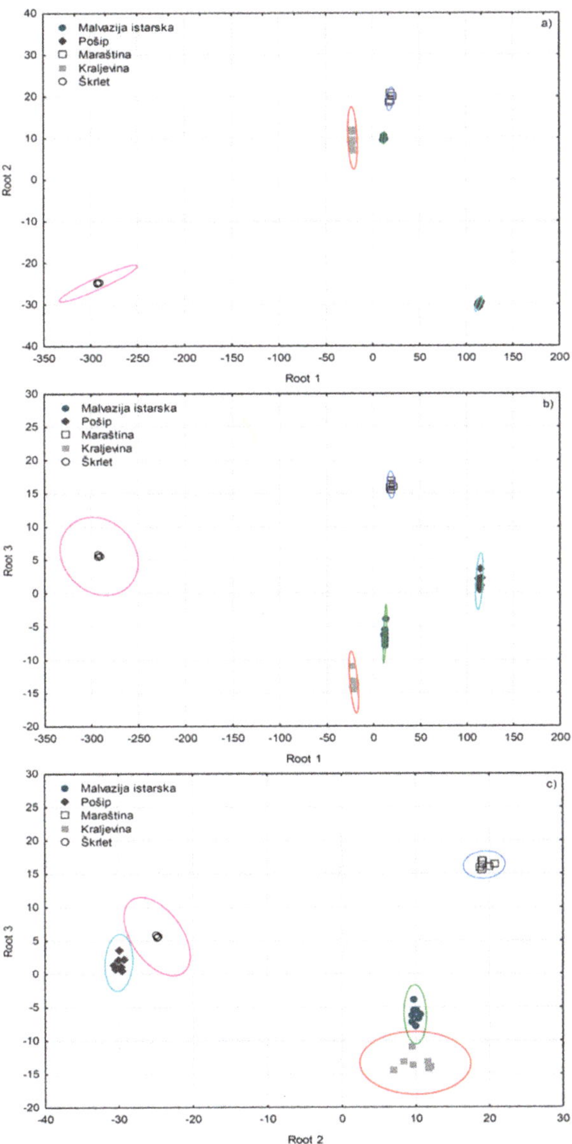

Figure 3. Separation of Croatian monovarietal wines according to variety defined by the first three discriminant functions (roots) obtained by forward stepwise discriminant analysis (SLDA) on the basis of volatile aroma compound composition determined by GC-MS and GC×GC-TOF-MS analysis. (**a**) root 1 vs root 2; (**b**) root 1 vs root 3; (**c**) root 2 vs root 3.

Apparently, SLDA has extracted volatile aroma compounds which were most useful for the differentiation of the five investigated monovarietal wines between each other, which only partly coincided with the compounds with the highest *F*-ratios obtained by ANOVA. Monoterpenes had a key role again, especially α-terpineol. The ability of the SLDA model to predict a correct variety was checked by "leave-one-out" cross-validation, where each wine sample was excluded and classified by the functions derived from all the other wine samples. The correct prediction rate achieved was 100.00%.

To compare the usefulness of the information contained in the composition of terpenes alone obtained by GC-MS and GC×GC-TOF-MS analysis for differentiating monovarietal wines, SLDA was applied separately on the two datasets containing 20 and 31 terpenes, respectively, found significant by ANOVA. Both GC-MS and GC×GC-TOF-MS dataset based models succeeded in achieving 100.00% correct classification (Figure 4). α-Terpineol was again confirmed as a key differentiator, since both models included it as the first, which classified correctly 59.38% and 68.75% monovarietal wines, respectively. For achieving 100.00% correct classification, the GC-MS model further included *trans*-ocimene, *cis*-linalool furan oxide, β-pinene, citronellol, *trans*-nerolidol, ho-trienol, *trans*-rose oxide, and limonene, while the GC×GC-TOF-MS model extracted γ-dehydro-ar-himachalene, ho-trienol, nerol, *o*-cymene, isogeraniol, a non-identified sesquiterpene (n.i.; *m/z* 119, 93, 69), neryl ethyl ether, and *cis*-α-ocimene. The classification efficacy of the models was improved by including further eight and nine terpenes, respectively. The GC×GC-TOF-MS model exhibited a superior efficacy judging from the degree of the overlapping of the corresponding 95% confidence areas, as well as higher squared Mahalanobis distances on the average, especially for Škrlet wines.

The volatile aroma compounds which were found to be most useful for the differentiation of the investigated wines in this study were only partly in accordance with the ones highlighted in previous studies which applied a similar multivariate statistical approach. For example, Welke et al. [29] characterized and differentiated wines from Chardonnay, Sauvignon Blanc, Pinot Noir, Merlot, and Cabernet Sauvignon based on volatile aroma composition obtained by GC×GC-TOF-MS analysis and extracted the following 12 volatile compounds as the most useful for their differentiation: diethyl succinate, 2,3-butanediol, nerol, 3-penten-2-one, diethyl malonate, β-santalol, ethyl 9-decenoate, alcohol-C_9, 4-carene, tetrahydro-2(2H)-pyranone, dihydro-2(3H)-thiophenone, and 3-methyl-2(5H)-furanone. It is probable that the main reason for such discrepancy between this and the study from Welke et al. [29] was the fact that the mentioned authors mutually compared wines from white and red varieties, which greatly differ with respect to the production technology, which, besides variety, certainly greatly contributed to the differences between wines. Welke et al. [29] also obtained a SLDA model that differentiated wines according to variety with a 100% correct recognition ability, while some other authors who applied conventional GC-MS for the same purpose, such as Zhang et al. [61] and Câmara, Alves and Marques [14], did not succeed completely. Fabani, Ravera, and Wunderlin [15] obtained a 100% correct discrimination among Syrah, Malbec, and Bonarda red wines by the application of SLDA on GC-MS data with ethyl hexanoate, ethyl octanoate, 1-hexanol, benzyl alcohol, and isoamyl acetate as the most useful differentiators. Terpenes were not analyzed. Ziółkowska, Wąsowicz, and Jeleń [19] obtained a relatively good differentiation of red wines, with the ability of the LDA model to correctly classify and predict their varietal origin based on HS-SPME/GC-MS data of 95%, while the model built for white wines was not that successful. The compounds most useful for the differentiation of white wines (Chardonnay, Sauvignon Blanc, and Muscat) were isoamyl acetate, furfural, ethyl octanoate, ethyl decanoate, and ethyl dodecanoate, while red wines (Cabernet Sauvignon and Merlot) were differentiated mainly by 1-hexanol, ethyl decanoate, and 2-phenylethanol. It should be noted that the samples of the same variety were collected across several countries, which was certainly a factor that introduced large variability.

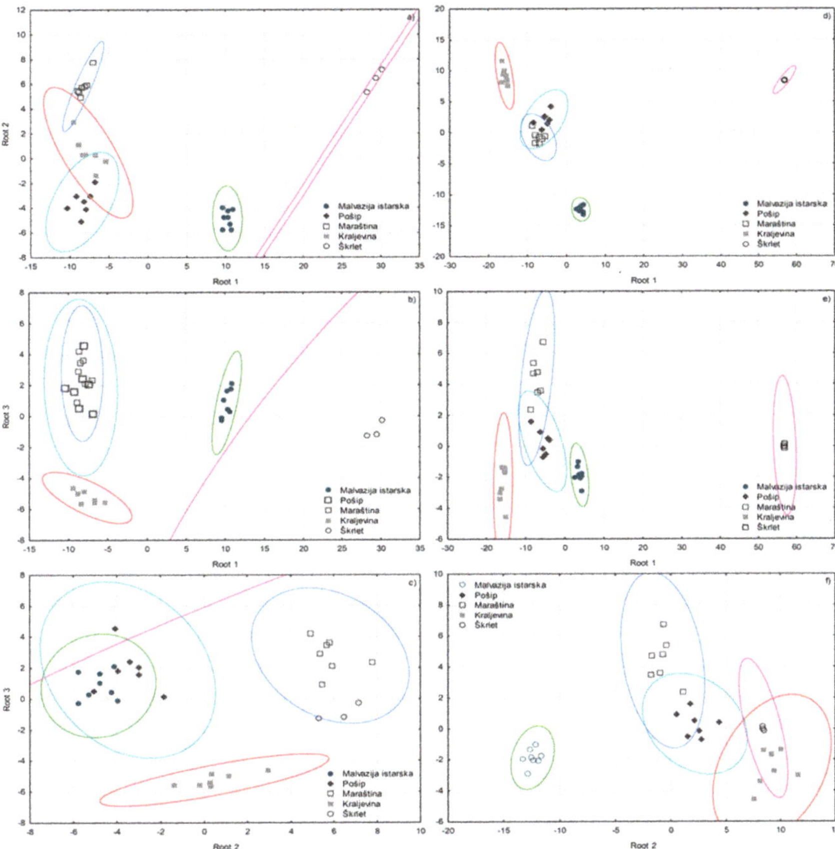

Figure 4. Separation of Croatian monovarietal wines according to variety defined by the first three discriminant functions (roots) obtained by forward stepwise discriminant analysis (SLDA) on the basis of volatile terpene composition determined by GC-MS (**a**–**c**) and GC×GC-TOF-MS (**d**–**f**) analysis.

4. Conclusions

HS-SPME/GC×GC-TOF-MS analysis, alone or combined with conventional HS-SPME/GC-MS, was shown to be an excellent analytical tool for differentiation of wines according to variety based on volatile aroma compound composition. It has also been proven that the additional separation efficiency enabled by the second chromatographic column in GC×GC-TOF-MS analysis was crucial for the separation and identification of a very large number of volatile compounds, which would otherwise remain undetected by conventional GC-MS. This feature provided additional in-depth volatile profile information which was exploited for highly efficient white wine varietal differentiation. Such an outcome can be considered even more successful knowing that the number of varieties was relatively high while that of wine samples of each variety was relatively small, and that the investigated wines were characterized by high intra-varietal heterogeneity in terms of micro-locations and grape cultivation and winemaking parameters. The results of this study confirmed the unmatched power of monoterpenes to discriminate wines according to variety, which was robust enough to be captured by uni- and multivariate statistics based on both GC-MS and GC×GC-TOF-MS analysis data separately.

Supplementary Materials: The following are available online at http://www.mdpi.com/2304-8158/9/12/1787/s1, Figure S1: Example of a contour plot obtained for Malvazija istarska monovarietal wine using HS-SPME/GC×GC-TOF-MS. Colored areas represent more abundant volatile aroma compounds and black dots represent less abundant and trace volatile aroma compounds, Table S1: Physico-chemical parameters in in Croatian monovarietal wines, Table S2: Concentrations (μg/L) of volatile aroma compounds found in individual Croatian monovarietal wines after headspace solid-phase microextraction followed by gas chromatography-mass spectrometry (HS-SPME/GC–MS) sorted by compound class, Table S3: Concentrations (μg/L relative to internal standard 2-octanol) of volatile aroma compounds found in individual Croatian monovarietal wines obtained by headspace solid-phase microextraction combined with comprehensive two-dimensional gas chromatography-mass spectrometry with time-of-flight mass spectrometric detection (HS-SPME/GC×GC-TOF-MS) sorted by compound class.

Author Contributions: Conceptualization, I.L.; methodology, I.L. and U.V.; formal analysis, I.L. and S.C.; investigation, I.L. and S.C.; resources, I.L. and U.V.; data curation, I.L. and U.V.; writing—original draft preparation, I.L.; writing—review and editing, I.L. and U.V.; supervision, I.L. and U.V.; project administration, I.L. and U.V.; funding acquisition, I.L. and U.V. All authors have read and agreed to the published version of the manuscript.

Funding: This research was funded by Croatian Science Foundation grant number UIP-2014-09-1194 and by the ADP 2017 project funded by the Autonomous Province of Trento.

Acknowledgments: The authors would like to thank wine producers from Croatia for donating wine samples, Irena Budić-Leto, Sanja Radeka, and the panel for sensory analysis of wine from the Institute of Agriculture and Tourism in Poreč, Croatia, for assistance in collecting and selection of wine samples, Ivana Puhelek for assistance in collecting wine samples, and Ivana Horvat for technical assistance.

Conflicts of Interest: The authors declare no conflict of interest. The funders had no role in the design of the study; in the collection, analyses, or interpretation of data; in the writing of the manuscript, or in the decision to publish the results.

References

1. Ribéreau-Gayon, P.; Glories, Y.; Maujean, A.; Dubourdieu, D. *Handbook of Enology, The Chemistry of Wine: Stabilization and Treatments*; John Wiley & Sons Ltd.: Chichester, UK, 2000; Volume 2, ISBN 9780471973638.
2. Rapp, A.; Mandery, H. Wine aroma. *Experientia* **1986**, *42*, 873–884. [CrossRef]
3. Ebeler, S.E. Analytical chemistry: Unlocking the secrets of wine flavor. *Food Rev. Int.* **2001**, *17*, 45–64. [CrossRef]
4. Ilc, T.; Werck-Reichhart, D.; Navrot, N. Meta-Analysis of the Core Aroma Components of Grape and Wine Aroma. *Front. Plant Sci.* **2016**, *7*, 1472:1–1472:15. [CrossRef]
5. Villano, C.; Lisanti, M.T.; Gambuti, A.; Vecchio, R.; Moio, L.; Frusciante, L.; Aversano, R.; Carputo, D. Wine varietal authentication based on phenolics, volatiles and DNA markers: State of the art, perspectives and drawbacks. *Food Control* **2017**, *80*, 1–10. [CrossRef]
6. Koundouras, S.; Marinos, V.; Gkoulioti, A.; Kotseridis, Y.; van Leeuwen, C.J. Influence of vineyard location and vine water status on fruit maturation of nonirrigated cv. Agiorgitiko (*Vitis vinifera* L.). Effects on wine phenolic and aroma components. *J. Agric. Food Chem.* **2006**, *54*, 5077–5086. [CrossRef]
7. Sabon, I.; de Revel, G.; Kotseridis, Y.; Bertrand, A. Determination of volatile compounds in Grenache wines in relation with different terroirs in the Rhone Valley. *J. Agric. Food Chem.* **2002**, *50*, 6341–6345. [CrossRef]
8. Yue, X.; Ma, X.; Tang, Y.; Wang, Y.; Wu, B.; Jiao, X.; Zhang, Z.; Ju, Y. Effect of cluster zone leaf removal on monoterpene profiles of Sauvignon Blanc grapes and wines. *Food Res. Int.* **2020**, *131*, 109028:1–109028:11. [CrossRef]
9. Šuklje, K.; Carlin, S.; Stanstrup, J.; Antalick, G.; Blackman, J.W.; Meeks, C.; Deloire, A.; Schmidtke, L.M.; Vrhovsek, U. Unravelling wine volatile evolution during shiraz grape ripening by untargeted HS-SPME-GC×GC-TOFMS. *Food Chem.* **2019**, *277*, 753–765. [CrossRef]
10. Falqué, E.; Fernández, E.; Dubourdieu, D. Differentiation of white wines by their aromatic index. *Talanta* **2001**, *54*, 271–281. [CrossRef]
11. Martínez-Pinilla, O.; Guadalupe, Z.; Ayestarán, B.; Pérez-Magariño, S.; Ortega-Heras, M. Characterization of volatile compounds and olfactory profile of red minority varietal wines from La Rioja. *J. Sci. Food Agric.* **2013**, *93*, 3720–3729. [CrossRef]
12. Barbará, J.A.; Primieri Nicolli, K.; Souza-Silva, É.A.; Telles Biasoto, A.C.; Welke, J.E.; Alcaraz Zini, C. Volatile profile and aroma potential of tropical Syrah wines elaborated in different maturation and maceration times using comprehensive two-dimensional gas chromatography and olfactometry. *Food Chem.* **2020**, *308*, 125552. [CrossRef]

13. Samoticha, J.; Wojdyło, A.; Chmielewska, J.; Nofer, J. Effect of Different Yeast Strains and Temperature of Fermentation on Basic Enological Parameters, Polyphenols and Volatile Compounds of Aurore White Wine. *Foods* **2019**, *8*, 599. [CrossRef]
14. Câmara, J.S.; Alves, M.A.; Marques, J.C. Multivariate analysis for the classification and differentiation of Madeira wines according to main grape varieties. *Talanta* **2006**, *68*, 1512–1521. [CrossRef]
15. Fabani, M.P.; Ravera, M.J.A.; Wunderlin, D.A. Markers of typical red wine varieties from the Valley of Tulum (San Juan-Argentina) based on VOCs profile and chemometrics. *Food Chem.* **2013**, *141*, 1055–1062. [CrossRef]
16. Gómez García-Carpintero, E.; Sánchez-Palomo, E.; Gómez Gallego, M.A.; González-Viñas, M.A. Free and bound volatile compounds as markers of aromatic typicalness of Moravia Dulce, Rojal and Tortosí red wines. *Food Chem.* **2012**, *131*, 90–98. [CrossRef]
17. Rocha, S.M.; Coutinho, P.; Coelho, E.; Barros, A.S.; Delgadillo, I.; Coimbra, M.A. Relationships between the varietal volatile composition of the musts and white wine aroma quality. A four year feasibility study. *LWT Food Sci. Technol.* **2010**, *43*, 1508–1516. [CrossRef]
18. Valentin, L.; Barroso, L.P.; Barbosa, R.M.; de Paulo, G.A.; Castro, I.A. Chemical typicality of South American red wines classified according to their volatile and phenolic compounds using multivariate analysis. *Food Chem.* **2020**, *302*, 125340:1–125340:8. [CrossRef]
19. Ziółkowska, A.; Wąsowicz, E.; Jeleń, H.H. Differentiation of wines according to grape variety and geographical origin based on volatiles profiling using SPME-MS and SPME-GC/MS methods. *Food Chem.* **2016**, *213*, 714–720. [CrossRef]
20. Welke, J.E.; Manfroi, V.; Zanus, M.; Lazarotto, M.; Alcaraz Zini, C. Characterization of the volatile profile of Brazilian Merlot wines through comprehensive two dimensional gas chromatography time-of-flight mass spectrometric detection. *J. Chromatogr. A* **2012**, *1226*, 124–139. [CrossRef]
21. Weldegergis, B.T.; de Villiers, A.; McNeish, C.; Seethapathy, S.; Mostafa, A.; Górecki, T.; Crouch, A.M. Characterisation of volatile components of Pinotage wines using comprehensive two-dimensional gas chromatography coupled to time-of-flight mass spectrometry (GC×GC–TOF-MS). *Food Chem.* **2011**, *129*, 188–199. [CrossRef]
22. Tranchida, P.Q.; Donato, P.; Cacciola, F.; Beccaria, M.; Dugo, P.; Mondello, L. Potential of comprehensive chromatography in food analysis. *TrAC Trend. Anal. Chem.* **2013**, *52*, 186–205. [CrossRef]
23. Purcaro, G.; Cordero, C.; Liberto, E.; Bicchi, C.; Conte, L.S. Toward a definition of blueprint of virgin olive oil by comprehensive two-dimensional gas chromatography. *J. Chromatogr. A* **2014**, *1334*, 101–111. [CrossRef] [PubMed]
24. Robinson, A.L.; Boss, P.K.; Heymann, H.; Solomon, P.S.; Trengove, R.D. Influence of Yeast Strain, Canopy Management, and Site on the Volatile Composition and Sensory Attributes of Cabernet Sauvignon Wines from Western Australia. *J. Agric. Food Chem.* **2011**, *59*, 3273–3284. [CrossRef]
25. Ryan, D.; Watkins, P.; Smith, J.; Allen, M.; Marriott, P. Analysis of methoxypyrazines in wine using headspace solid phase microextraction with isotope dilution and comprehensive two-dimensional gas chromatography. *J. Sep. Sci.* **2005**, *28*, 1075–1082. [CrossRef]
26. Šuklje, K.; Carlin, S.; Antalick, G.; Blackman, J.W.; Deloire, A.; Vrhovsek, U.; Schmidtke, L.M. Regional Discrimination of Australian Shiraz Wine Volatome by Two-Dimensional Gas Chromatography Coupled to Time-of-Flight Mass Spectrometry. *J. Agric. Food Chem.* **2019**, *67*, 10273–10284. [CrossRef]
27. Welke, J.E.; Zanus, M.; Lazzarotto, M.; Alcaraz Zini, C. Quantitative analysis of headspace volatile compounds using comprehensive two-dimensional gas chromatography and their contribution to the aroma of Chardonnay wine. *Food Res. Int.* **2014**, *59*, 85–99. [CrossRef]
28. Carlin, S.; Vrhovsek, U.; Lonardi, A.; Landi, L.; Mattivi, F. Aromatic complexity in Verdicchio wines: A case study. *OENO One* **2019**, *4*, 597–610. [CrossRef]
29. Welke, J.E.; Manfroi, V.; Zanus, M.; Lazzarotto, M.; Alcaraz Zini, C. Differentiation of wines according to grape variety using multivariate analysis of comprehensive two-dimensional gas chromatography with time-of-flight mass spectrometric detection dana. *Food Chem.* **2013**, *141*, 3897–3905. [CrossRef]
30. Bubola, M.; Lukić, I.; Radeka, S.; Sivilotti, P.; Grozić, K.; Vanzo, A.; Bavčar, D.; Lisjak, K. Enhancement of Istrian Malvasia wine aroma and hydroxycinnamate composition by hand and mechanical leaf removal. *J. Sci. Food Agric.* **2019**, *99*, 904–914. [CrossRef]

31. Carlin, S.; Vrhovsek, U.; Franceschi, P.; Lotti, C.; Bontempo, L.; Camin, F.; Toubiana, D.; Zottele, F.; Toller, G.; Fait, A.; et al. Regional features of northern Italian sparkling wines, identified using solid-phase micro extraction and comprehensive two-dimensional gas chromatography coupled with time-of-flight mass spectrometry. *Food Chem.* **2016**, *208*, 68–80. [CrossRef]
32. Beckner Whitener, M.E.; Stanstrup, J.; Panzeri, V.; Carlin, S.; Divol, B.; Toit, M.; Vrhovsek, U. Untangling the wine metabolome by combining untargeted SPME–GCxGC-TOF-MS and sensory analysis to profile Sauvignon blanc co-fermented with seven different yeasts. *Metabolomics* **2016**, *12*, 53:1–53:25. [CrossRef]
33. Xia, J.; Sinelnikov, I.; Han, B.; Wishart, D.S. MetaboAnalyst 3.0 – Making metabolomics more meaningful. *Nucleic Acids Res.* **2015**, *43*, W251–W257. [CrossRef]
34. Black, C.A.; Parker, M.; Siebert, T.E.; Capone, D.L.; Francis, I.L. Terpenoids and their role in wine flavour: Recent advances. *Aust. J. Grape Wine R* **2015**, *21*, 582–600. [CrossRef]
35. Ribéreau-Gayon, P.; Glories, Y.; Maujean, A.; Dubourdieu, D. *Handbook of Enology, The Chemistry of Wine: Stabilization and Treatments*, 2nd ed.; John Wiley & Sons: Chichester, UK, 2006; Volume 2, pp. 205–230. ISBN 978-0-470-01037-2.
36. Mateo, J.J.; Jiménez, M. Monoterpenes in grape juice and wines. *J. Chromatogr. A* **2000**, *881*, 557–567. [CrossRef]
37. Guth, H. Quantitation and sensory studies of character impact odorants of different white wine varieties. *J. Agric. Food Chem.* **1997**, *45*, 3027–3032. [CrossRef]
38. Battilana, J.; Emanuelli, F.; Gambino, G.; Gribaudo, I.; Gasperi, F.; Boss, P.K.; Grando, M.S. Functional effect of grapevine 1-deoxy-D-xylulose 5-phosphate synthase substitution K284N on Muscat flavour formation. *J. Exp. Bot.* **2011**, *62*, 5497–5508. [CrossRef]
39. Duchêne, E.; Legras, J.L.; Karst, F.; Merdinoglu, D.; Claudel, P.; Jaegli, N.; Pelsy, F. Variation of linalool and geraniol content within two pairs of aromatic and non-aromatic grapevine clones. *Aust. J. Grape Wine R.* **2009**, *15*, 120–130. [CrossRef]
40. Skinkis, P.A.; Bordelon, B.P.; Wood, K. Comparison of Monoterpene Constituents in Traminette, Gewürztraminer, and Riesling Winegrapes. *Am. J. Enol. Vitic.* **2008**, *59*, 440–445.
41. Song, M.; Fuentes, C.; Loos, A.; Tomasino, E. Free Monoterpene Isomer Profiles of *Vitis vinifera* L. cv. White Wines. *Foods* **2018**, *7*, 27. [CrossRef]
42. Versini, G.; Orriols, I.; Dalla Serra, A. Aroma components of Galician Albariño, Loureira and Godello wines. *Vitis* **1994**, *33*, 165–170. [CrossRef]
43. Lukić, I.; Horvat, I. Differentiation of Commercial PDO Wines Produced in Istria (Croatia) According to Variety and Harvest Year Based on HS-SPME-GC/MS Volatile Aroma Compound Profiling. *Food Technol. Biotechnol.* **2017**, *55*, 95–108. [CrossRef]
44. Câmara, J.S.; Alves, M.A.; Carlos Marques, J.C. Classification of Boal, Malvazia, Sercial and Verdelho wines based on terpenoid patterns. *Food Chem.* **2007**, *101*, 475–484. [CrossRef]
45. Falqué, E.; Fernández, E.; Dubourdieu, D. Volatile Components of Loureira, Dona Branca, and Treixadura Wines. *J. Agric. Food Chem.* **2002**, *50*, 538–543. [CrossRef]
46. Márquez, R.; Castro, R.; Natera, R.; García-Barroso, C. Characterisation of the volatile fraction of Andalusian sweet wines. *Eur. Food. Res. Technol.* **2008**, *226*, 1479–1484. [CrossRef]
47. Radeka, S.; Herjavec, S.; Peršurić, Đ.; Lukić, I.; Sladonja, B. Effect of different maceration treatments on free and bound varietal aroma compounds in wine of *Vitis vinifera* L. cv. Malvazija istarska bijela. *Food Technol. Biotechnol.* **2008**, *46*, 86–92.
48. Lukić, I.; Jedrejčić, N.; Kovačević Ganić, K.; Staver, M.; Peršurić, Đ. Phenolic and aroma composition of white wines produced by prolonged maceration and maturation in wooden barrels. *Food Technol. Biotechnol.* **2015**, *53*, 407–418. [CrossRef]
49. Lukić, I.; Horvat, I.; Radeka, S.; Damijanić, K.; Staver, M. Effect of different levels of skin disruption and contact with oxygen during grape processing on phenols, volatile aromas, and sensory caracteristics of white wine. *J. Food Process. Preserv.* **2019**, *43*, e13969:1–e13969:12. [CrossRef]
50. González-Barreiro, C.; Rial-Otero, R.; Cancho-Grande, B.; Simal-Gándara, J. Wine aroma compounds in grapes: A critical review. *Crit. Rev. Food Sci.* **2013**, *55*, 202–228. [CrossRef]
51. Tonietto, J.; Carbonneau, A. A multicriteria climatic classification system for grape-growing regions worldwide. *Agr. Forest Meteorol.* **2004**, *124*, 81–97. [CrossRef]

52. Belancic, A.; Agosin, E.; Ibacache, A.; Bordeu, E.; Baumes, R.; Razungles, A.; Bayonove, C. Influence of sun exposure on the aromatic composition of chilean Muscat grape cultivars Moscatel de Alejandria and Moscatel rosada. *Am. J. Enol. Vitic.* **1997**, *48*, 181–186.
53. Mendes-Pinto, M.M. Carotenoid breakdown products - the norisoprenoids - in wine aroma. *Arch. Biochem. Biophys.* **2009**, *483*, 236–245. [CrossRef]
54. Marais, J.; van Wyk, C.J.; Rapp, A. Effect of sunlight and shade on norisoprenoids levels in maturing Weisser Riesling and Chenin blanc grapes and Weisser Riesling wines. *S. Afr. J. Enol. Vitic.* **1992**, *13*, 23–32. [CrossRef]
55. Oliveira, J.M.; Faria, M.; Sá, F.; Barros, F.; Araújo, I.M. C6-alcohols as varietal markers for assessment of wine origin. *Anal. Chim. Acta* **2006**, *563*, 300–309. [CrossRef]
56. Ugliano, M.; Henschke, P.A. Yeasts and wine flavour. In *Wine Chemistry and Biochemistry*; Moreno-Arribas, M.V., Polo, M.C., Eds.; Springer: New York, NY, USA, 2009; pp. 313–392. ISBN 978-0-387-74116-1.
57. Louw, L.; Tredoux, A.G.J.; Van Rensburg, P.; Kidd, M.; Naes, T.; Nieuwoudt, H.H. Fermentation derived aroma compounds in varietal young wines from South Africa. *S. Afr. J. Enol. Vitic.* **2010**, *31*, 213–225. [CrossRef]
58. Ferreira, V.; López, R.; Cacho, J.F. Quantitative determination of the odorants of young red wines from different grape varieties. *J. Sci. Food Agric.* **2002**, *80*, 1659–1667. [CrossRef]
59. Schreier, P.; Jennings, W.G. Flavor composition of wines: A review. *Crit. Rev. Food Sci. Nutr.* **1979**, *12*, 59–111. [CrossRef]
60. Verstrepen, K.J.; Van Laere, S.D.M.; Vanderhaegen, B.M.P.; Derdelinckx, G.; Dufour, J.-P.; Pretorius, I.S.; Winderickx, J.; Thevelein, J.M.; Delvaux, F.R. Expression levels of the yeast alcohol acetyltransferase genes ATF1, Lg-ATF1, and ATF2 control the formation of a broad range of volatile esters. *Appl. Environ. Microbiol.* **2003**, *69*, 5228–5237. [CrossRef]
61. Zhang, J.; Li, L.; Gao, N.; Wang, D.; Gao, Q.; Jiang, S. Feature extraction and selection from volatile compounds for analytical classification of Chinese red wines from different varieties. *Anal. Chim. Acta* **2010**, *662*, 137–142. [CrossRef]

Article

Comprehensive Chemical and Sensory Assessment of Wines Made from White Grapes of *Vitis vinifera* Cultivars Albillo Dorado and Montonera del Casar: A Comparative Study with Airén

José Pérez-Navarro [1], Pedro Miguel Izquierdo-Cañas [2,3], Adela Mena-Morales [2], Juan Luis Chacón-Vozmediano [2], Jesús Martínez-Gascueña [2], Esteban García-Romero [2], Isidro Hermosín-Gutiérrez [1,†] and Sergio Gómez-Alonso [1,*]

1 Regional Institute for Applied Scientific Research (IRICA), University of Castilla-La Mancha, Av. Camilo José Cela, 10, 13071 Ciudad Real, Spain; jose.pnavarro@uclm.es

2 Instituto Regional de Investigación y Desarrollo Agroalimentario y Forestal de Castilla-La Mancha (IRIAF), Ctra. Albacete s/n, 13700 Tomelloso, Spain; pmizquierdo@jccm.es (P.M.I.-C.); amenam@jccm.es (A.M.-M.); jlchacon@jccm.es (J.L.C.-V.); jmartinezg@jccm.es (J.M.-G.); estebang@jccm.es (E.G.-R.)

3 Parque Científico y Tecnológico de Castilla-La Mancha, Paseo de la Innovación 1, 02006 Albacete, Spain

* Correspondence: sergio.gomez@uclm.es

† In memoriam.

Received: 17 August 2020; Accepted: 10 September 2020; Published: 12 September 2020

Abstract: The ability to obtain different wines with a singular organoleptic profile is one of the main factors for the wine industry's growth, in order to appeal to a broad cross section of consumers. Due to this, white wines made from the novel grape genotypes Albillo Dorado and Montonera del Casar (*Vitis vinifera* L.) were studied and compared to the well-known Airén at two consecutive years. Wines were evaluated by physicochemical, spectrophotometric, high-performance liquid chromatography–diode array detection–mass spectrometry, gas chromatography–mass spectrometry and sensory analyses. The chromatic characteristics of the new wines were defined by more color purity than Airén, with greenish highlights. In general, the phenolic profile of the Albillo Dorado wines showed a higher flavonol and hydroxycinnamic acid derivative content. Several volatile compounds were determined, and their odor activity values were calculated to determine their impact on wine aroma. A fruity series dominated the wine aromatic composition, but spicier and greener notes characterized the aroma profile of Airén wines. Albillo Dorado and Montonera del Casar were sensory evaluated as wines with a less fresh taste compared to Airén. Unique chemical and sensory profiles were determined for wines made from these novel grape genotypes, providing alternative monovarietal wines to encourage the wine market growth and extend the offer to consumers.

Keywords: white wines; color; phenolic composition; volatile aroma compounds; odor activity values; sensory profile

---

1. Introduction

Dominated by the main international grape cultivars, the development of different white wine types is mostly determined by wine consumer preferences and needs in the global market [1]. White wines with specific organoleptic characteristics have been requested in the last years, being the most appreciated those that are characterized by varietal aromas, with green and tropical fruit notes and an elegant acid taste [2]. Wine sensory attributes are defined by their physical and chemical composition, mainly phenolic and volatile compounds, which play an important role in wine quality. White wines typically show a lower content of phenolic compounds than red ones [3]. Their polyphenolic fraction is usually dominated by hydroxycinnamic acid derivatives, hydroxybenzoic acids, flavonols and flavan-3-ols; it has been demonstrated that these compounds are related to wine sensorial properties, such as the chromatic characteristics, color stability, bitterness and astringency [4]. Regarding volatile compounds, terpenes and norisoprenoids are associated with grape cultivars and may be important for the expression of wine varietal characteristics [5].

Different styles of white wine elaboration have been developed in the last years, with the purpose to diversify the wine market. Some alternative practices to obtain wines characterized by singular sensorial properties are the fermentation and ageing in a barrique on lees [6], fermentation by endogenous yeast microflora [7] or prolonged maceration during and after fermentation and ageing in wooden barrels [8]. The use of indigenous or novel grape cultivars is another feasible option to make singular white wines. Recently, there is an increase of research works about indigenous grape cultivars from different countries such as Greece, Italy, Japan and Spain [9–12].

In south-central Spain, Castilla-La Mancha is the biggest wine area in the world, with 470,000 ha of vineyard. Conformed by several tens of indigenous grape cultivars and a significant number of foreign ones, the vine heritage of this region is going through a period of impoverishment due to the inclusion of foreign cultivars that may result in the autochthonous grape extinction. Because of the plant material loss, Instituto Regional de Investigación y Desarrollo Agroalimentario y Forestal (IRIAF) of Castilla-La Mancha carried out a vineyard prospective study to recover and characterize the indigenous grape cultivars of this region. That work allowed the identification of more than 40 new grape genotypes not previously described in the literature consulted [13,14]. The Grapevine Germplasm Bank of Castilla-La Mancha (GGBCM) was created to preserve the great diversity of grape varieties from this region, where they grow at optimal virus-free conditions. Albillo Dorado and Montonera del Casar are two novel white grape genotypes (*Vitis vinifera* L.) recently identified in Castilla-La Mancha, and they were found in different areas of this region: Albillo Dorado grapes are located in Albacete, and the Montonera del Casar genotype is distributed in several zones, i.e., Cuenca, Guadalajara and Toledo. Recently, the flavanol glycoside content of the grape seeds and skins of these novel genotypes was studied [15]. In addition, wines made from several minor grapes have been previously studied in Castilla-La Mancha, whose cultivation is restricted to small areas. For example, the volatile and sensory profile of wines elaborated with the Moravia Agria grape cultivar and the phenolic composition of Moravia Dulce, Rojal and Tortosí wines have been reported in the literature [16,17]. However, there is no known data about the physicochemical characteristics of wines made from these two novel grape genotypes.

The knowledge of the oenological potential of alternative grape cultivars can give opportunities to obtain distinctive wines that appeal to a broad cross section of consumers. Therefore, the aim of this work was to perform a detailed characterization of young white wines made from the novel grape genotypes Albillo Dorado and Montonera del Casar (*Vitis vinifera* L.) recently identified in Castilla-La Mancha, at two consecutive years (2015 and 2016). The study included the comparison to Airén, currently the most cultivated grape variety in this region and one of the most important white grapes in the world. Wines were assessed by physicochemical, spectrophotometric, high-performance liquid chromatography–diode array detection–electrospray ionization-mass spectrometry (HPLC-DAD-ESI-MS/MS), gas chromatography–mass spectrometry (GC-MS) and sensory analyses.

2. Materials and Methods

2.1. Chemicals

Analytical grade chemicals (>99%) and Milli-Q water (Merk-Millipore, Darmstadt, Germany) were used. The analysis of phenolic compounds was performed using solvents of LC-MS quality (Fisher Scientific, Madrid, Spain): acetonitrile (CH_3CN), formic acid (HCOOH) and methanol (CH_3OH). The following GC-MS grade solvents were used for the volatile compound analysis:

n-pentane ($CH_3CH_2CH_2CH_2CH_3$) and dichloromethane (CH_2Cl_2) that were purchased from Scharlab (Sentmenat, Spain), and Fisher Scientific (Madrid, Spain) provided the ethyl acetate ($CH_3COOCH_2CH_3$) and CH_3OH. Phenolic compound identification and quantitation were carried out using commercial standards from Phytolab (Vestenbergsgreuth, Germany): (-)-epigallocatechin, (-)-gallocatechin, quercetin 3-glucuronide, *trans*-caftaric acid and *trans*-piceid. Procyanidins B1, B2 and B4, quercetin 3-galactoside, quercetin 3-rutinoside, kaempferol 3-galactoside, kaempferol 3-glucuronide, kaempferol 3-rutinoside, the 3-glucoside derivatives of quercetin, kaempferol, isorhamnetin and their aglycones were supplied by Extrasynthese (Genay, France). Moreover, (-)-epicatechin, (-)-epicatechin 3-gallate, (+)-catechin and *trans*-resveratrol commercial standards were used from Sigma-Aldrich (Tres Cantos, Madrid, Spain). *Cis*-resveratrol and *cis*-piceid were obtained from their *trans* isomers subjected to UV-irradiation (366 nm light for 5 min in quartz vials) in a 25% CH_3OH solution. For identifying and quantifying the volatile compounds, several chemical standards were purchased from Extrasynthese (Genay, France), Fluka (Buchs, Germany), Sigma-Aldrich (Steinheim, Germany) and Merck (Darmstard, Germany). The 4-nonanol and 4-methyl-2-pentanol compounds were obtained from Sigma-Aldrich (Steinheim, Germany), both used as internal standards.

2.2. Winemaking

Albillo Dorado, Montonera del Casar and Airén grapes were cultivated in the same vineyard located in the GGBCM under the warm climate of this Spanish region. Grapes were harvested in two consecutive years (2015 and 2016), at the optimal ripening stage for winemaking: sugar content, 19.3–22.5° Brix; total acidity, 2.7–5.6 g L^{-1}; and pH, 3.5–3.9. The winemaking was made in duplicate, employing 75 kg of each grape genotype to elaborate monovarietal white wines at the experimental winery of IRIAF. When grapes were destemmed and crushed, the addition of SO_2 (80 mg L^{-1}) protected the grape must from undesirable microorganisms. A cold pre-fermentative maceration was carried out at 5 °C during 24 h. The alcoholic fermentation started in stainless steel tanks (100 L capacity) by inoculation with *Saccharomyces cerevisiae* yeast (Uvaferm VN®, Lallemand Inc., Zug, Switzerland) at 20 g hL^{-1}. The fermentation was conducted at 17 °C, controlled by density measure and finished when the glucose + fructose level was <5 g L^{-1}. Afterwards, the wines were racked and adjusted to 25 mg L^{-1} SO_2. The Battonnage technique was performed for one month, which consists of stirring the dregs regularly. Finally, the wines were stabilized at −5 °C for 15 days, corrected up to 25 mg L^{-1} SO_2, passed through 0.2 μm filters, bottled and stored at 16–18 °C. The wines were elaborated and analyzed in their vintage years.

2.3. Wine Physicochemical and Spectrophotometric Analyses

The analytical methods proposed by the International Organisation of Vine and Wine (OIV) [18] were used to determine the following wine physicochemical parameters: alcoholic strength, relative density, total and volatile acidity, pH, glucose + fructose concentration, glycerin content, SO_2 and the concentration of different acids. White wines were subjected to CIELab color space through the MSCV® software [19] in order to obtain their chromatic characteristics. Total phenolic content was analyzed according to the method proposed by the Folin–Ciocalteu method [20] and a precipitation assay with methyl-cellulose was followed for the condensed tannin measurements [21].

2.4. HPLC-DAD-ESI-MS/MS Analysis of Flavonols and Hydroxycinnamic Acid Derivatives

The flavonols and hydroxycinnamic acid derivatives (HCADs) were analyzed on an Agilent 1100 Series HPLC system (Agilent, Waldbronn, Germany), equipped with a diode array detector (model G1315B) and mass spectrometry detector with an electrospray ionization source (LC/MSD Trap VL, model G2445C VL). Before analysis, 2 mL of the wine were dried in a rotary evaporator at 35 °C and reconstituted in 1 mL of 20% CH_3OH. The sample was injected (40 μL) on a reversed-phase column ZORBAX Eclipse XDB-C18 (2.1 × 150 mm; 3.5 μm particle, Agilent USA, Santa Clara-California) at 40 °C. The chromatographic conditions used were as follows: Solvent A

($CH_3CN/H_2O/HCOOH$, 3:88.5:8.5 *v/v/v*), Solvent B ($CH_3CN/H_2O/HCOOH$, 50:41.5:8.5, *v/v/v*) and Solvent C ($CH_3OH/H_2O/HCOOH$, 90:1.5:8.5, *v/v/v*), with a flow rate of 0.19 mL min^{-1}. For the HPLC analysis of these phenolic compounds, the solvent gradient was (time, % each solvent) zero min, 96% A and 4% B; 8 min, 96% A and 4% B; 37 min, 70% A, 17% B and 13% C; 51 min, 50% A, 30% B and 20% C; 51.5 min, 30% A, 40% B and 30% C; 56 min, 50% B and 50% C; 57 min, 50% B and 50% C; and 64 min, 96% A and 4% B. For identification purposes, the mass spectrometer was operated under the following conditions: negative ionization mode; scan range, 100–1000 *m/z*; dry gas, N_2, flow 8 L min^{-1}; drying temperature, 350 °C; nebulizer, 40 psi; capillary, +3500 V; capillary exit offset, −68 V; skimmer 1, −20 V; skimmer 2, −60 V. The identification of these compounds was based on their UV-Vis and MS/MS spectra, obtained from standards and matches such as those reported in the literature [22,23]. Flavonol quantitation was performed using DAD-chromatograms extracted at 360 nm. The total concentration of flavonols was expressed as equivalents of quercetin 3-glucoside. For the hydroxycinnamic acid derivatives, DAD-chromatograms at 320 nm were used and their total concentration was expressed as *trans*-caftaric acid equivalents (Table S1).

2.5. HPLC-MS/MS-MRM Analysis of Flavan-3-ols and Stilbenes

Flavan-3-ol and the stilbene fractions were obtained by solid phase extraction (SPE) using C18 cartridges (Sep-pak Plus C18, 820 mg of adsorbent, Waters Corporation, Milford, USA). The extraction and analysis of these compounds were carried out following the procedure described in the literature [24]. Separation, identification and quantitation of wine flavan-3-ols and stilbenes were performed on an HPLC Agilent 1200 series system (Agilent, Waldbronn, Germany) coupled to a DAD (Agilent, Waldbronn, Germany) and AB Sciex 3200 QTRAP triple quadrupole mass spectrometer with turbo spray ionization (Applied Biosystems, Foster City, CA, USA) (ESI–MS/MS) in the Multiple Reaction Monitoring (MRM) mode. Analyst MSD software (Applied Biosystems, version 1.5) was used to process the mass spectral data. The total flavan-3-ol monomer concentration was expressed as (+)-catechin equivalents, flavan-3-ol dimers were quantified as equivalents of procyanidin B1 and stilbene concentration was expressed as *trans*-resveratrol equivalents.

2.6. GC-MS Analysis of Volatile Aroma Compounds

Volatile compound extraction was obtained following the methodology reported in [25], using SPE cartridges (LiChrolut EN, Merck, 0.3 g of phase, Darmstadt, Germany) and an internal standard (4-nonanol, 0.1 g L^{-1}). A volume of 25 mL of wine, containing the internal standard, were passed through preconditioned cartridges and then the resin was washed with 25 mL of Milli-Q water in order to remove the sugars and polar compounds. The volatile fraction elution was carried out with 15 mL of pentane-dichloromethane (2:1, *v/v*). Then, the extracts were concentrated to 150 μL by distillation in a Vigreux column and then under a nitrogen stream and kept at −20 °C until GC-MS analysis. Wine volatile compounds were determined using a gas chromatograph FocusGC coupled to a mass spectrometer ISQ with electron impact ionization source and quadrupole analyzer, and equipped with an autosampler TriPlus (ThermoQuest, Waltham, MA, USA). A capillary column BP21 (SGE, Ringwood, Australia) (50 m × 0.32 mm internal diameter; 0.25 μm film thickness; FFPA stationary phase, polyethylene glycol treated with nitroterephthalic acid) was used. For the analysis of major volatile compounds, 100 μL of wine was diluted with 100 μL of 4-methyl-2-pentanol (50 mg L^{-1}) as the internal standard and 1 mL of Milli-Q water [26]. The sample was injected (0.8 μL) in split mode (split ratio: 10) at 195 °C. Helium was used as the carrier gas with a constant flow of 1.2 mL min^{-1}. The oven temperature was 32 °C during 2 min, 5 °C min^{-1} to 120 °C, 75 °C min^{-1} to 190 °C and 18 min at 190 °C. Regarding the minor volatile compounds, 1 μL of the SPE eluate was injected in the splitless mode (splitless time: 0.3 min). The chromatographic conditions were as follows: oven temperature, 40 °C for 15 min, 2 °C min^{-1} to 100 °C, 1 °C min^{-1} to 150 °C, 4 °C min^{-1} to 210 °C and 55 min at 210 °C; injector temperature, 220 °C; and carrier helium gas, 1 mL min^{-1}. The following detector parameters were set: mass scanning range, 40–250 amu; ion source temperature 250 °C;

impact energy, 70 eV; and electron multiplier voltage, 1603 V. The mass-spectral library, retention times and pure commercial volatile compounds allowed the volatile compound identification. The specific *m/z* fragment for each compound was used for the quantification by GC-MS, using the internal standard method. The concentration of non-available commercial compounds was expressed as equivalents of the internal standard obtained by normalizing the peak area of each compound to that of the internal standard and multiplying by the internal standard concentration [27].

2.7. Wine Sensory Analysis

White wines made from the novel grape genotypes and Airén were sensory evaluated by 11 panelists with considerable experience in tasting wines (staff members from IRIAF trained in descriptive sensory analysis of wines, age range from 25 to 65, 60% female and 40% male), under ISO standards related to methodology and vocabulary, tasting room and taster selection and training [28–30]. The Napping® technique was performed to determine the organoleptic properties of wines [31]. According to similarities or differences in the sensory profile, the wine samples were placed on a white paper sheet of 40 cm × 60 cm without imposing structure or attributes, which allowed to identify the most important attributes for tasters. The results obtained from each wine sample were two coordinates (X and Y) that can be converted into a distance matrix. A further session was planned with a list of wine attributes imposed by the chief judge and previously chosen by an expert panel. The distribution of the wine samples on the blank sheet was established according to similarities and dissimilarities between wines with the attributes defined. Once the evaluation sessions finished, data obtained from each taster and wine sample provided the coordinates and frequency tables.

2.8. Statistical Analysis

All chemical data were treated using a one-way analysis of variance (ANOVA, Student–Newman–Keuls test, $p < 0.05$) to identify significant differences between the wines between the two studied years. Statistical tests were performed using SPSS software, version 23.0 (IBM, Endicott, New York, USA). Sensorial data were processed with XLSTAT 2017 statistical software (Addinsoft, Paris, France).

3. Results and Discussion

3.1. Physicochemical Parameters and Chromatic Characteristics

Table S2 shows the physiochemical data of white wines made from Airén and the novel grape genotypes. Alcoholic fermentation finished correctly as indicated by the low content of fructose and glucose (≤0.35 g L^{-1}), being considered as dry white wines (reducing sugars < 5 g L^{-1}). The optimal total acidity (3.7–5.8 g L^{-1}), pH (3.4–3.6) and volatile acidity (0.2–0.4 g L^{-1}) values were determined for all wines, with a suitable alcoholic content and within the usual parameters shown by white wines made in this Spanish region [32].

Regarding the chromatic characteristics, no significant differences were found between all evaluated wines for L*, which corresponded to lightness. The Montonera del Casar and Albillo Dorado wines provided greater color purity than the Airén ones (chroma, C*). Moreover, the parameter a* and b* measured for the new white wines differed from that observed in Airén, with higher b* values and more negatives for a*. This was manifested by a more yellow color with greenish highlights, an appealing feature for this type of wines.

The use of novel grape genotypes did not affect the total polyphenol content of the wines since no statistical differences were measured, accounting 300–410 mg L^{-1} as gallic acid equivalents. The new wines were characterized by a significant lower condensed tannin concentration than the Airén ones. Those data do not offer much information about the phenolic compound influence on the interesting sensory properties of the wines. Due to this, the detailed composition of several phenolic compounds classes was studied.

3.2. Phenolic Composition

The flavonol profile of the wines assessed is shown in Figure S1. The 3-glycoside series of quercetin aglycone were determined for all wines, comprising 3-glucoside, 3-galactoside and 3-glucuronide. In addition, the 3-glucoside and 3-galactoside derivatives of kaempferol and isorhamnetin were also identified, and the free aglycones released from them by hydrolysis in wine. Results obtained were in accordance with reported data for white grapes and wines, with the presence of the flavonols mono- and di-substituted in the B-ring [33]. The wine analysis showed the absence of the B-ring tri-substituted flavonols, which are myricetin, laricitrin and syringetin. As expected in white wines, the profile of the wines made from Airén and the novel grape genotypes was dominated by quercetin-type flavonols, with a proportion of 80–100% (Table 1). Kaempferol-based flavonols were the second more important ones found in the new wines (7–16%) and absent in the Airén ones. The Albillo Dorado wines were characterized by the exclusive presence of kaempferol- and isorhamnetin 3-galactosides.

Moreover, the flavonol fraction of the Airén and Albillo Dorado wines showed a compound with a molecular ion at *m/z* 447 in the MS spectrum that suffered the 146 amu loss in the ion trap, which results in a product ion at *m/z* 301, corresponding to the deprotonated moiety of quercetin aglycone. This allowed the identification of the compound quercetin 3-rhamnoside, previously reported in wines made from Moribel grapes [23]. A high degree of flavonol hydrolysis was determined in the wines, reaching 80% in Montonera del Casar. However, the flavonols were not hydrolyzed in the 2015 Airén wines, which can be explained by the low content of these compounds. The most abundant aglycone was quercetin, which is more polar and soluble in a hydro-alcoholic solution like wine. The total content of flavonols measured by HPLC accounted 0.1–11.4 mg L^{-1} as quercetin 3-glucoside equivalents. These values were low due to the winemaking process, since flavonols are mainly located in grape skins and their concentration in wines is dependent on skin extraction. Besides the abovementioned, the flavonol concentration of the studied wines was lower compared to other grape variety wines, such as Macabeo white wines from this region [34]. Although wine flavonols could be affected by many factors [35,36], grape variety also introduced variability in the total concentration of these compounds since all the studied wines were elaborated using the same conditions and winemaking techniques, and Albillo Dorado provided the flavonol-richest wines both years.

The monomeric flavan-3-ols determined in the wines are shown in Table S3. The (+)-catechin compound accounted for approximately half of the wine monomers, followed by (-)-epicatechin and (-)-gallocatechin. In addition, the glycosylated derivatives of the flavan-3-ol monomers were quantified in all the wines. Montonera del Casar wines were characterized by the lowest monomeric flavan-3-ol concentration (ca. 1.7 mg L^{-1} as (+)-catechin equivalents) and Albillo Dorado with values close to Airén (16 mg L^{-1}) in 2016. The dimeric flavan-3-ol profile of the wines was dominated by procyanidin B1, with a higher proportion in the 2016 Montonera del Casar wines. A greater proportion of procyanidin B4 was found in the Airén wines and the galloylated dimers in the Albillo Dorado ones. Wines made from Airén grapes were characterized by the highest concentration of flavan-3-ol dimers, similar to monomers. These compounds and their polymers contribute to a mouthfeel sensation of white wines, such as bitterness and astringency [37].

Table 1. Chromatographic and spectroscopic characteristics, molar proportions and total concentration (mean value ± standard deviation, $n = 2$) of the flavonols and hydroxycinnamic acid derivatives identified in white wines made from Airén and the novel grape genotypes.

Peak [1]	RT (min)	Compounds	Molecular and Product Ions (m/z)	Airén 2015	Airén 2016	Albillo Dorado 2015	Albillo Dorado 2016	Montonera del Casar 2015	Montonera del Casar 2016
		Flavonols (mg L^{-1}) [2]		0.14 ± 0.01 a	1.70 ± 0.74 a	4.53 ± 0.34 b	11.40 ± 1.60 c	1.93 ± 0.45 a	0.55 ± 0.02 a
1	26.21	Quercetin 3-galactoside	463, 301	ND	9.89 ± 2.73 b	7.08 ± 2.04 ab	5.55 ± 0.37 ab	4.60 ± 0.97 a	ND
2	26.74	Quercetin 3-glucuronide	477, 301	38.49 ± 9.56 b	ND	ND	ND	4.50 ± 2.14 a	4.64 ± 0.64 a
3	28.23	Quercetin 3-glucoside	463, 301	15.09 ± 1.48 ab	11.98 ± 3.74 ab	22.05 ± 0.19 c	17.41 ± 1.18 b	9.41 ± 0.82 a	12.57 ± 0.25 ab
4	31.95	Kaempferol 3-galactoside	447, 285	ND	ND	2.45 ± 0.46 b	1.57 ± 0.00 a	ND	ND
5	32.66	Quercetin 3-rhamnoside	447, 301	46.42 ± 8.08 b	3.72 ± 1.40 a	0.68 ± 0.05 a	0.60 ± 0.02 a	ND	ND
6	34.98	Kaempferol 3-glucoside	447, 285	ND	ND	2.56 ± 0.60 a	2.64 ± 0.19 a	ND	ND
7	36.65	Isorhamnetin 3-galactoside	477, 315	ND	ND	0.50 ± 0.17 a	0.37 ± 0.05 a	ND	ND
8	38.24	Isorhamnetin 3-glucoside	477, 315	ND	1.47 ± 0.12 a	2.29 ± 1.17 b	1.51 ± 0.11 a	1.26 ± 0.35 a	ND
9	42.98	Quercetin	301	ND	72.94 ± 7.74 b	52.02 ± 1.48 a	56.93 ± 1.96 a	72.98 ± 4.49 b	82.79 ± 0.89 b
10	51.68	Kaempferol	285	ND	ND	8.40 ± 2.48 a	11.34 ± 0.14 b	7.26 ± 0.15 a	ND
11	55.31	Isorhamnetin	315	ND	ND	1.96 ± 0.96 a	2.08 ± 0.70 a	ND	ND
		% Quercetin-type		100.00 ± 0.00 e	98.53 ± 0.12 d	81.83 ± 0.33 b	80.49 ± 0.43 a	91.48 ± 0.20 c	100.00 ± 0.00 e
		% Kaempferol-type		ND	ND	13.42 ± 2.63 b	15.56 ± 0.33 b	7.26 ± 0.15 a	ND
		% Isorhamnetin-type		ND	1.47 ± 0.12 a	4.76 ± 2.30 c	3.96 ± 0.76 b	1.26 ± 0.35 a	ND
		% Hydrolysis		00.00 ± 0.00 a	72.94 ± 7.74 bc	62.38 ± 3.00 b	70.36 ± 1.39 bc	80.24 ± 4.64 c	82.79 ± 0.89 c
		Hydroxycinnamic Acid Derivatives (mg L^{-1}) [3]		50.53 ± 2.47 b	42.04 ± 0.97 a	66.05 ± 0.55 c	54.06 ± 1.42 b	41.45 ± 2.04 a	40.58 ± 1.96 a
12	2.15	Caffeoyl-glucose	341, 179, 161	11.11 ± 1.84 a	13.82 ± 1.23 ab	16.56 ± 3.08 bc	19.07 ± 0.49 c	30.15 ± 0.83 d	32.13 ± 0.68 e
13	3.46	*trans*-2-S-Glutathionyl-caftaric acid	616, 484, 440, 272	13.29 ± 0.45 a	13.43 ± 0.74 a	15.01 ± 3.89 a	17.62 ± 0.78 a	19.97 ± 1.13 a	46.19 ± 1.79 b
14	3.84	*trans*-Caftaric acid	311, 179, 149	42.22 ± 0.25 cd	37.48 ± 0.06 c	42.85 ± 4.87 d	32.60 ± 0.29 bc	29.20 ± 2.87 b	6.71 ± 0.43 a
15	4.43	*cis*-2-S-Glutathionyl-caftaric acid	616, 484, 440, 272	0.44 ± 0.07 a	1.27 ± 0.02 b	0.90 ± 0.18 ab	1.01 ± 0.21 b	0.87 ± 0.22 ab	1.25 ± 0.15 b
16	5.58	*trans*-Coutaric acid	295, 163, 149	11.62 ± 0.10 b	12.28 ± 0.27 b	9.86 ± 2.55 b	12.37 ± 0.55 b	3.69 ± 0.24 a	2.14 ± 0.31 a
17	5.74	*cis*-Coutaric acid	295, 163, 149	6.04 ± 0.39 c	5.12 ± 0.02 bc	4.76 ± 0.56 b	5.44 ± 0.39 bc	1.69 ± 0.42 a	ND
18	6.88	*p*-Coumaroyl-glucose	325, 163, 145	5.28 ± 0.07 e	4.51 ± 0.15 d	2.30 ± 0.28 b	1.03 ± 0.05 a	3.59 ± 0.39 c	1.30 ± 0.35 a
19	7.95	*trans*-Fertaric acid	325, 193, 149	7.15 ± 0.48 b	9.92 ± 0.45 d	6.07 ± 0.04 a	8.85 ± 0.23 c	8.48 ± 0.39 c	8.58 ± 0.01 c
20	8.60	*cis*-Fertaric acid	325, 193, 149	1.44 ± 0.08 a	1.51 ± 0.03 a	1.07 ± 0.30 a	1.50 ± 0.06 a	1.30 ± 0.22 a	1.35 ± 0.18 a
21	43.74	Ethyl caffeate	207, 179, 135	1.41 ± 0.05 c	0.68 ± 0.10 ab	0.62 ± 0.35 ab	0.50 ± 0.17 a	1.08 ± 0.04 bc	0.35 ± 0.06 a
		% 2-S-Glutathionyl-caftaric acid		13.73 ± 0.52 a	14.70 ± 0.72 ab	15.92 ± 4.06 ab	18.63 ± 0.57 ab	20.83 ± 0.91 b	47.44 ± 1.65 c
		% Caffeic-type		54.74 ± 1.65 b	51.98 ± 1.07 b	60.02 ± 2.14 c	52.17 ± 0.62 b	60.43 ± 2.08 c	39.19 ± 1.17 a
		% *p*-Coumaric-type		22.94 ± 0.57 d	21.90 ± 0.13 d	16.92 ± 2.27 c	18.85 ± 0.89 c	8.96 ± 0.57 b	3.44 ± 0.66 a
		% Ferulic-type		8.59 ± 0.56 b	11.43 ± 0.48 d	7.14 ± 0.35 a	10.35 ± 0.30 c	9.78 ± 0.61 bc	9.93 ± 0.19 bc

[1] Peak number used in Figures S1 and S2. [2] As quercetin 3-glucoside equivalents. [3] As *trans*-caftaric acid equivalents. Abbreviations: ND, not detected. Different letters in the same row indicate that the values are significantly different (ANOVA, Student–Newman–Keuls test, $p < 0.05$).

Regarding the non-flavonoid compounds, hydroxycinnamic acids exist in wine as free form or esterified with tartaric acid. The expected hydroxycinnamoyl-tartaric acids (caftaric, coutaric and fertaric acids) were the hydroxycinnamic acid derivatives (HCADs) found in all wines, and also the ethyl ester of caffeic acid. Figure S2 shows the profile of these compounds in wines made from Airén and the novel grape genotypes. Corresponding to fertaric acid, the extracted ion chromatogram *m/z* 325 was also detected with different retention time and fragmentation pattern. This compound was a glucoside derivative of p-coumaric acid, previously reported in wines made from BRS Violeta grapes [22]. The identification of 2-S-glutathionylcaftaric acid (grape reaction product or GRP) was confirmed by UV-vis and MS/MS spectra. This compound is a reaction product of glutathione with oxidized caftaric acid, and its *trans* and *cis* isomers were found in all the wines. The formation of GRP avoids the browning of wines and also provides information of the oxidation suffered by the wine during its elaboration and aging [38]. With regards to the HCAD-types, the analyzed wines had greater caffeic-type proportions (39–60%), mainly *trans*-caftaric acid, except in the case of Montonera del Casar wines (Table 1). Statistical differences were observed in the total content of hydroxycinnamic acid derivatives among the wines. The highest HCAD concentration was determined in the Albillo Dorado wines—in both years. Data obtained in 2015 were in agreement with those reported for Chardonnay wines from Castilla-La Mancha [39].

Table S3 shows the stilbenes based on resveratrol identified in the studied wines. The main stilbenes were *cis*-resveratrol (32–62%) and *cis*-piceid, its 3-glucoside derivative (29–45%). Montonera del Casar provided wines with the lowest total stilbene concentrations, accounting around 0.08 mg L^{-1} as equivalents of *trans*-resveratrol and similar to those reported for *Vitis vinifera* white wines [40]. Nevertheless, these values were approximately half the concentrations in wines made from Albillo Dorado and Airén grapes (0.13–0.32 mg L^{-1}).

3.3. Volatile Composition and Aromatic Series

Volatile compounds are crucial to the quality of wines due to their influence on the wine aroma profile. Concentrations of the most relevant volatile compounds on white wine aroma, their odor descriptors, aromatic series and thresholds are shown in Table 2. These volatile compounds belonged to different chemical families, e.g., acids, alcohols, aldehydes, benzenic compounds, C6 compounds, esters, furanic compounds, lactones, norisoprenoids and terpenes. Some of them have originated from grapes (varietal aromas) or were produced during alcoholic fermentation (fermentative aromas).

C6 compounds form part of the wine varietal aroma and are related to green and herbaceous nuances. They can provide undesirable flavors at high concentrations, having a negative impact on wine quality [41]. These compounds are considered one of the most important groups of varietal aromas in the wines assessed due to their concentrations. The main six carbon alcohol was *cis*-3-hexenol in wines made from Airén and Albillo Dorado grapes, accounting for 0.6–1.9 mg L^{-1}. A lower concentration of its *trans* isomer was determined, in accordance with reported data for white wines [34]. However, the ratio between *trans*- and *cis*-3-hexenol moved in the opposite direction to Montonera del Casar. Wines were characterized by a higher total C6 compound content than other white wines elaborated with grapes grown in this region [39], and Airén exhibited the richest profile of these compounds. Other typical varietal aroma compounds are terpenes that provide citric and floral aromas [42]. Citronellol, geraniol and linalool were the main terpenes determined in all the wines. With values ranging between 2.9 and 6.8 μg L^{-1}, the concentration of geraniol was generally the biggest and close to data reported for Chardonnay wines from Castilla-La Mancha [43]. In this work, terpene concentrations were under their threshold values, so the contribution of these compounds to wine aroma did not seem significant. Norisoprenoids are volatile compounds related to the varietal typicity of wines with low odor thresholds [42,44]. β-damascenone provided fruity aromas since its concentration exceeded the odor threshold value (0.05 μg L^{-1}) [42] in all wines. Although there were no significant differences in total concentration of benzenic compounds between wines, several individual compounds differed between each other. The benzenic compound profile of Montonera del Casar wines showed higher guaiacol concentration in 2016 (11.8 μg L^{-1}), with a value above its odor threshold (10 μg L^{-1}) [45]. Moreover, it was characterized by having the greatest concentration of benzaldehyde in both years. The content of this compound has not exceeded its olfactory threshold (2000 μg L^{-1}) [46] but it could have a synergic effect to wine aroma, contributing to fruity and floral notes. Related to sweet spice aroma, eugenol was found at a higher concentration in Airén wines. The content of this compound and vanillin did not exceed their odor threshold values [47].

During alcoholic fermentation, alcohols are formed as a byproduct of yeast amino acid metabolism. In isolation, these compounds do not typically have desirable odors and could be a negative factor on wine quality if their concentration is above 400 mg L^{-1} [48]. Since the total concentration of these compounds was under this value (284–345 mg L^{-1}), they can make a positive contribution to wine complexity. Due to their higher solubility and volatility, mid-change fatty acids can have an important impact on wine flavor, providing fruity, cheese, fatty, and rancid notes. The studied wines had acid contents within the range 14–76 mg L^{-1}, showing higher values in 2016 and with similar 6, 8 and 10-carbon fatty acid concentrations to those reported for Muscat wines [49]. The main esters found in wines were the ethyl esters and acetates of fatty acids. These compounds are key contributors to the fruity aromas of wines [50] and produced by yeasts in an enzymatic process during alcoholic fermentation. The new wines were characterized by higher ester concentrations than the Airén ones in 2015. Acetaldehyde is a wine aldehyde associated with fruity and nutty aromas and formed mainly by yeast metabolism during alcoholic fermentation, before ethanol formation. The wine acetaldehyde concentration depends on the yeast strain and initial must SO_2 concentration; consequently, the higher acetaldehyde content in Airén wines can be due to the must composition since all the wines were made under the same conditions.

Table 2. Aroma volatile composition (mean value ± standard deviation, $n = 2$) in white wines made from Airén and the novel grape genotypes.

Compounds	m/z	Odor Descriptors	Odorant Series [1]	Odor Threshold (µg L^{-1})	Airén		Albillo Dorado		Montonera del Casar	
					2015	2016	2015	2016	2015	2016
		C6 Compounds [2]			2.24 ± 0.00 bc	2.83 ± 0.59 c	1.12 ± 0.06 a	1.82 ± 0.09 b	0.55 ± 0.05 a	0.52 ± 0.01 a
1-Hexanol [2]	69	Flower, green, cut grass	II, III	8000 [42]	0.54 ± 0.01 a	0.80 ± 0.18 b	0.39 ± 0.02 a	0.66 ± 0.03 ab	0.42 ± 0.04 a	0.39 ± 0.01 a
trans-3-Hexenol [2]	67	Green	III	400 [46]	0.19 ± 0.00 b	0.17 ± 0.04 b	0.11 ± 0.03 a	0.16 ± 0.00 b	0.07 ± 0.01 a	0.08 ± 0.00 a
cis-3-Hexenol [2]	67	Green, cut grass	III	400 [42]	1.51 ± 0.01 d	1.86 ± 0.37 d	0.62 ± 0.01 b	1.01 ± 0.01 c	0.06 ± 0.00 a	0.05 ± 0.00 a
		Terpenes [3]			10.15 ± 2.93 a	10.02 ± 1.78 a	10.19 ± 1.14 a	9.11 ± 1.14 a	10.37 ± 1.16 a	12.41 ± 0.66 a
Linalool [3]	93	Floral	II	15 [42]	1.80 ± 0.05 a	2.98 ± 0.58 b	1.65 ± 0.22 a	2.78 ± 0.20 b	3.51 ± 0.54 b	3.17 ± 0.19 b
Citronellol [3]	95	Floral	II	100 [42]	3.20 ± 1.06 a	2.11 ± 0.29 a	4.02 ± 0.78 a	2.21 ± 0.35 a	3.98 ± 1.71 a	2.49 ± 0.04 a
Geraniol [3]	93	Roses, geranium	II	30 [42]	5.16 ± 1.83 a	4.94 ± 0.91 a	4.53 ± 0.14 a	4.12 ± 0.60 a	2.88 ± 1.08 a	6.75 ± 0.43 a
Norisoprenoids [3]					3.87 ± 0.12 a	3.60 ± 0.54 a	6.75 ± 1.10 b	10.66 ± 3.00 ab	5.56 ± 0.55 ab	7.78 ± 2.67 ab
β-Damascenone [3]	190	Sweet, fruit	I, IV	0.05 [42]	3.84 ± 0.13 a	3.50 ± 0.52 a	6.59 ± 1.11 ab	10.57 ± 2.98 b	5.29 ± 0.33 ab	7.76 ± 2.67 ab
β-Ionone [3]	177	Floral, violet	II	0.09 [45]	0.03 ± 0.01 a	0.10 ± 0.02 a	0.16 ± 0.00 a	0.08 ± 0.01 a	0.27 ± 0.13 a	0.02 ± 0.00 a
		Benzenic Compounds [2]			20.58 ± 0.11 a	17.71 ± 5.85 a	24.62 ± 6.67 a	17.87 ± 1.19 a	14.85 ± 1.82 a	11.80 ± 0.29 a
Benzaldehyde [3]	105	Sweet, cherry, almond	I, IV	2000 [46]	2.62 ± 0.08 b	2.32 ± 0.79 b	1.03 ± 0.17 a	0.98 ± 0.15 a	13.73 ± 0.11 c	44.13 ± 0.31 d
Eugenol [3]	164	Spices, clove, honey	IV, V, VII	6 [45]	2.14 ± 0.24 b	2.86 ± 0.15 c	0.83 ± 0.31 a	0.93 ± 0.17 a	0.43 ± 0.07 a	0.82 ± 0.18 a
Guaiacol [3]	124	Medicine, caramel, smoke	IV, VI	10 [45]	9.18 ± 2.40 b	2.14 ± 0.47 a	4.08 ± 0.02 a	2.30 ± 0.66 a	8.71 ± 2.42 b	11.82 ± 1.43 b
Vanillin [3]	151	Vanilla	V, VII	60 [47]	7.24 ± 1.88 a	7.33 ± 1.26 a	7.26 ± 1.31 a	9.44 ± 2.51 a	8.17 ± 1.34 a	5.75 ± 0.40 a
2-Phenylethanol [2]	122	Floral, roses	II	10,000 [42]	20.56 ± 0.11 a	17.70 ± 5.85 a	24.60 ± 6.67 a	17.86 ± 1.19 a	14.82 ± 1.81 a	11.74 ± 0.29 a
		Alcohols [2]			313.62 ± 14.63 a	284.26 ± 2.73 a	327.07 ± 49.62 a	324.12 ± 34.58 a	307.10 ± 6.56 a	345.45 ± 1.27 a
Methanol [2]	31	Chemical, medicinal	VI	668,000 [46]	31.05 ± 4.91 a	48.95 ± 1.90 b	44.77 ± 1.11 b	40.53 ± 1.24 ab	76.14 ± 7.74 c	120.37 ± 3.28 d
Isoamyl alcohol [2]	55	Solvent, fusel	VII	30,000 [42]	228.31 ± 6.31 a	200.17 ± 0.75 a	227.24 ± 43.99 a	231.18 ± 21.49 a	182.94 ± 3.48 a	173.87 ± 0.71 a
3-Ethoxy-propanol [3]	59	Fruity	I	100 [51]	8.10 ± 0.24 a	3.55 ± 1.63 a	3.58 ± 0.29 a	2.22 ± 1.59 a	24.77 ± 3.38 c	14.58 ± 1.90 b
3-Methylthio-propanol [2]	106	Cooked, vegetable	VI	500 [42]	0.44 ± 0.02 b	0.25 ± 0.08 ab	0.09 ± 0.03 a	0.22 ± 0.10 a	0.26 ± 0.00 ab	0.25 ± 0.04 ab
Isobutanol [2]	74	Bitter, green	III, VI	40,000 [42]	53.82 ± 3.37 a	34.88 ± 3.54 a	54.97 ± 6.70 a	52.19 ± 11.76 a	47.74 ± 2.31 a	50.95 ± 1.34 a
		Acids [2]			19.04 ± 1.10 a	74.61 ± 8.93 b	14.37 ± 0.23 a	63.56 ± 1.47 b	21.96 ± 0.20 a	75.58 ± 9.13 b
Butyric acid [2]	73	Rancid, cheese, sweat	VI	173 [45]	0.42 ± 0.03 ab	0.49 ± 0.07 ab	0.08 ± 0.02 a	0.32 ± 0.09 ab	0.70 ± 0.29 b	0.58 ± 0.04 b
Isovaleric acid [2]	87	Acid, rancid	IV, VI	33 [45]	1.07 ± 0.05 a	1.08 ± 0.30 a	0.45 ± 0.17 a	0.81 ± 0.22 a	0.74 ± 0.06 a	0.91 ± 0.11 a
Hexanoic acid [2]	60	Sweat	VI	420 [45]	7.23 ± 0.01 ab	12.29 ± 1.83 c	4.99 ± 0.01 a	10.87 ± 0.52 b	7.42 ± 1.34 ab	10.16 ± 0.39 bc
Octanoic acid [2]	60	Sweat, cheese	VI	500 [45]	7.02 ± 0.87 a	48.48 ± 7.08 b	5.92 ± 0.11 a	38.65 ± 1.28 b	7.72 ± 0.59 a	43.93 ± 4.87 b
Decanoic acid [2]	73	Rancid fat	VI	1000 [45]	3.31 ± 0.20 a	12.26 ± 0.35 b	2.91 ± 0.19 a	12.91 ± 0.38 b	5.37 ± 0.91 a	19.99 ± 3.71 c
		Aldehydes [2]			60.93 ± 9.36 c	76.06 ± 6.43 d	52.11 ± 2.21 bc	41.43 ± 3.46 ab	31.77 ± 6.23 a	32.17 ± 3.58 a
Acetaldehyde [2]	43	Pungent, ripe apple	I, VI	500 [42]	60.93 ± 9.36 c	76.06 ± 6.43 d	52.11 ± 2.21 bc	41.43 ± 3.46 ab	31.77 ± 6.23 a	32.17 ± 3.58 a

Table 2. *Cont.*

					Airén		Albillo Dorado		Montonera del Casar	
		Esters [2]			91.36 ± 0.45 b	58.75 ± 5.45 a	107.65 ± 12.96 c	67.73 ± 5.70 a	111.22 ± 4.60 c	76.24 ± 3.01 ab
Ethyl acetate [2]	70	Fruity, solvent	I, VI	7500 [42]	62.43 ± 1.94 ab	43.35 ± 5.13 a	69.37 ± 14.74 ab	50.21 ± 5.49 ab	73.13 ± 5.40 b	60.28 ± 2.18 ab
Isoamyl acetate [2]	61	Banana	I	30 [45]	17.31 ± 0.56 c	7.08 ± 1.13 a	11.89 ± 0.81 b	6.97 ± 0.60 a	16.19 ± 1.79 c	6.32 ± 0.41 a
2-Phenylethyl acetate [2]	104	Floral, roses	II	250 [42]	0.25 ± 0.00 a	0.26 ± 0.03 a	0.25 ± 0.01 a	0.26 ± 0.02 a	0.20 ± 0.01 a	0.26 ± 0.02 a
Ethyl lactate [2]	75	Acid, medicine	VI	154,636 [47]	8.84 ± 2.06 a	5.06 ± 0.52 a	23.75 ± 2.88 b	7.41 ± 0.72 a	19.24 ± 2.67 b	6.47 ± 0.44 a
Ethyl butyrate [2]	88	Fruity	I	20 [42]	0.62 ± 0.01 b	0.25 ± 0.06 a	0.69 ± 0.11 b	0.30 ± 0.04 a	0.86 ± 0.06 c	0.26 ± 0.01 a
Ethyl hexanoate [2]	88	Green apple	I	14 [45]	0.70 ± 0.00 a	1.03 ± 0.12 b	0.57 ± 0.04 a	0.89 ± 0.06 b	0.53 ± 0.00 a	0.85 ± 0.00 b
Ethyl octanoate [2]	88	Sweet, fruity	I, II, IV	5 [45]	1.06 ± 0.01 abc	1.47 ± 0.19 c	0.98 ± 0.13 ab	1.34 ± 0.17 bc	0.92 ± 0.00 a	1.39 ± 0.04 c
Ethyl decanoate [2]	88	Sweet, fruity	I, IV	200 [45]	0.17 ± 0.00 a	0.27 ± 0.01 b	0.15 ± 0.01 a	0.36 ± 0.07 c	0.18 ± 0.02 a	0.41 ± 0.01 c
		Furanic Compounds [3]			13.22 ± 0.07 ab	27.34 ± 4.71 b	3.24 ± 2.12 a	15.24 ± 8.36 ab	7.50 ± 1.95 a	17.40 ± 3.93 ab
Furaneol [3]	128	Burnt sugar, caramel	IV	5 [52]	13.22 ± 0.07 ab	27.34 ± 4.71 b	3.24 ± 2.12 a	15.24 ± 8.36 ab	7.50 ± 1.95 a	17.40 ± 3.93 ab
		Lactones [3]			58.92 ± 4.11 a	149.15 ± 37.65 b	51.00 ± 1.98 a	144.67 ± 39.34 b	75.53 ± 12.25 a	175.37 ± 3.37 b
γ-Butyrolactone [3]	86	Sweet, toast, caramel	IV	35,000 [53]	42.76 ± 3.35 a	37.74 ± 8.96 a	39.82 ± 3.60 a	66.94 ± 20.94 a	57.28 ± 2.69 a	106.08 ± 3.57 b
γ-Nonalactone [3]	85	Coconut	IV	30 [45]	1.52 ± 0.14 a	3.29 ± 0.68 b	2.68 ± 0.25 b	7.19 ± 0.50 d	4.88 ± 0.58 c	6.29 ± 0.13 d
δ-Dodecalactone [3]	114	Coconut	IV	88 [46]	14.64 ± 0.90 a	108.12 ± 28.01 c	8.50 ± 1.87 a	70.54 ± 17.90 b	13.37 ± 8.98 a	63.00 ± 0.07 b

[1] I = fruity; II = floral; III = green, fresh; IV = sweet; V = spicy; VI = fatty; VII = others. [2] mg L^{-1}. [3] μg L^{-1}. Different letters in the same row indicate that the values are significantly different (ANOVA, Student–Newman–Keuls test, $p < 0.05$).

Among the main lactones determined in these white wines, γ-butyrolactone showed the highest concentrations with values below its odor threshold (35 mg L^{-1}) [53]. Other lactones identified, such as γ-nonalactone and δ-dodecalactone, are also found in beer so they may be fermentation products.

Each volatile compound identified has a different impact on the overall aroma character of wines. To assess the influence of the aroma compounds in the studied wines, the Odor Activity Values (OAVs) were determined by dividing the compound concentration in the wine by the concentration corresponding to its odor threshold from the literature [42,45–47,51–53]. To estimate the overall wine aroma, the following aromatic series was established to bring together the volatile compounds according to their odor descriptors: green/fresh, fatty, floral, fruity, spicy, sweet and other odors. The values of the aromatic series were calculated based on summation of the OAVs for volatile compounds assigned to each series. The total intensity of each aromatic series (ΣOAVs) used to define the aroma profile of the white wines studied is shown in Figure 1. The highest aroma contribution to the wines was the fruity series that was the most characteristic attribute of the wine aromatic profile, followed by the sweet series. There are statistically significant differences in the intensity values of the green/fresh and spicy series between the wines, Airén showing the highest ones in both years. This could be explained by its C6 compound and eugenol concentrations, related to higher intensities of fresh and spice notes, respectively. The floral and green/fresh series were used by winetasters to define the sensory aroma profile of these wines although they are not the most abundant aromatic series. This issue can be attributed to the fact that the intensity of the sensory attributes of the wine aroma is affected by suppression, synergy and effects on the wine matrix, which is not taking into account in the OAV determination.

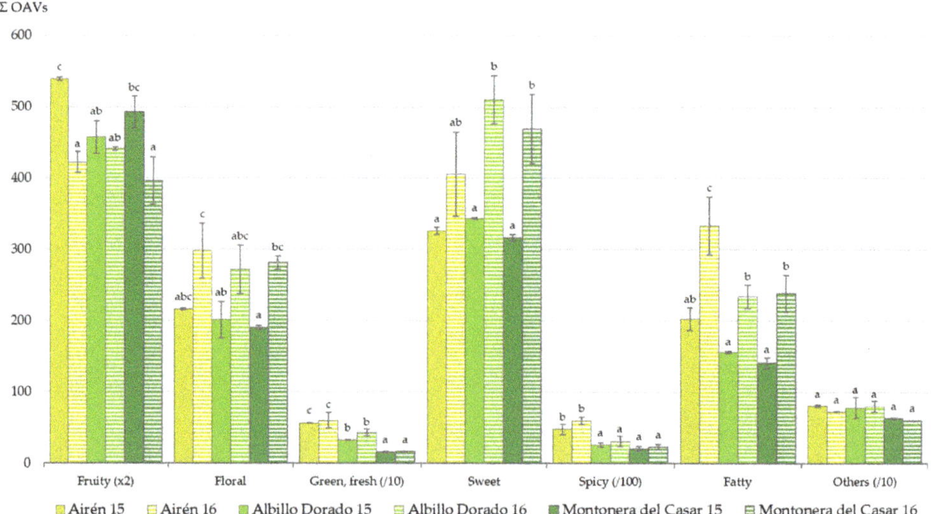

Figure 1. Aromatic series (ΣOAVs, mean value ± standard deviation, $n = 2$) in white wines made from Airén and the novel grape genotypes at two consecutive years (2015 and 2016). Different letters in the same aromatic series indicate that the values are significantly different among the wines (ANOVA, Student–Newman–Keuls test, $p < 0.05$).

3.4. Sensory Profile

The sensory characterization of the wines made from Airén and the novel grape genotypes was performed using the Napping® technique by experienced winetasters, professionals with great knowledge of wines. A simplified view of the sensory characteristics of the wines on a two-dimensional map is shown in Figure 2. The first two dimensions accounted for 66.19% of the explained variance

(48.28% and 17.91%, respectively). According to the results, it was observed that each monovarietal wine produced from the two years, 2015 and 2016, appear close together, which indicates only little sensorial differences among the vintages. In addition, all wines were placed on the right side of the graph, showing that the wines were described by a somewhat similar sensory profile. However, each wine was defined by singular organoleptic properties. Airén wines were characterized by green apple notes, a distinguishing attribute in wines made from this grape variety [43]. Floral aromas with banana notes defined the sensory profile of the Albillo Dorado wines, similar to its homonym, the Albillo grape cultivar [49]. A herbaceous odor descriptor was correlated with the profile of Montonera del Casar wines. Concerning mouthfeel properties, wines made from the novel grape genotypes presented a less fresh taste and acidity than the Airén ones. The sensory assessment underlined the differences among the wines according to grape genotype, impacting on the global organoleptic quality in accordance with the chemical composition.

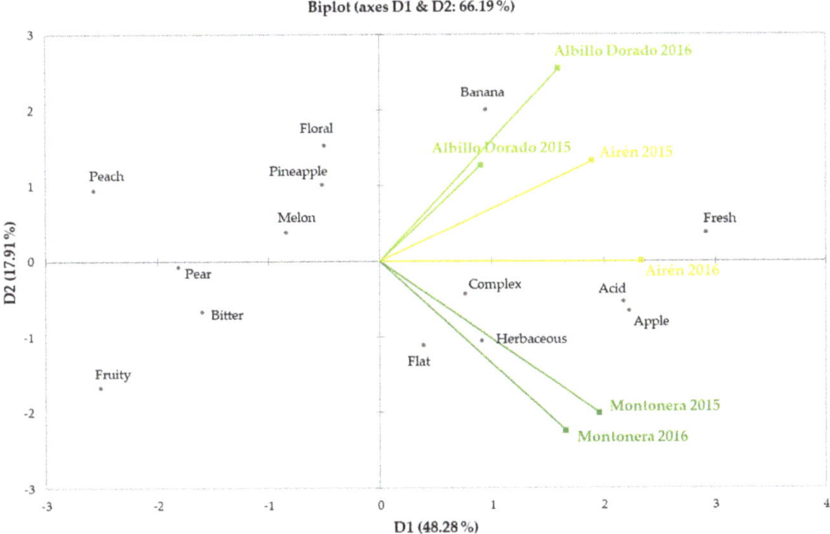

Figure 2. Sensory characterization of the white wines made from Airén and the novel grape genotypes at two consecutive years (2015 and 2016).

4. Conclusions

An exhaustive characterization of young white wines made from the novel grape genotypes Albillo Dorado and Montonera del Casar (*Vitis vinifera* L.) was performed in two consecutive years, comparing them to wines made from the Airén cultivar. Several significant differences in terms of chromatic characteristics and chemical composition (phenolic and volatile compounds) were established for the wines assessed. In addition, each wine was sensory characterized by singular organoleptic properties, with the positive and desirable attributes for these types of wines.

The data presented herein provide valuable information about the oenological potential of these novel grape genotypes for the first time. Moreover, the findings could be useful for the winemaking sector, mainly wineries and producers, in order to obtain distinctive monovarietal wines from these grape genotypes, increasing the diversification of white wines in the market and extending the offer to a broad cross-section of consumers.

Supplementary Materials: The following are available online at http://www.mdpi.com/2304-8158/9/9/1282/s1, Figure S1: Flavonol profiles (HPLC DAD-chromatograms at 360 nm) in white wines made from Airén and novel grape genotypes: (a) Airén, (b) Albillo Dorado, (c) Montonera del Casar. Peak numbering is as in Table 1, Figure S2: Hydroxycinnamic acid derivative profiles (HPLC DAD-chromatograms at 320 nm) in white wines made from Airén and novel grape genotypes: (a) Airén, (b) Albillo Dorado, (c) Montonera del Casar. Peak numbering is as in Table 1, Table S1: Calibration data of commercial standards used for the quantification of flavonols and hydroxycinnamic acid derivatives in white wines made from Airén and novel grape genotypes, Table S2: Physicochemical parameters, global phenolic composition and chromatic characteristics (mean value ± standard deviation, $n = 2$) in white wines made from Airén and novel grape genotypes, Table S3: Flavan-3-ol monomer, dimer and stilbene molar percentage and total concentration (mean value ± standard deviation, $n = 2$) in white wines made from Airén and novel grape genotypes.

Author Contributions: Conceptualization, P.M.I.-C., E.G.-R., I.H.-G. and S.G.-A.; methodology, J.P.-N., P.M.I.-C., A.M.-M., J.L.C.-V., J.M.-G., E.G.-R., I.H.-G. and S.G.-A.; formal analysis, J.P.-N., P.M.I.-C., A.M.-M.; investigation, J.P.-N., P.M.I.-C., A.M.-M., J.L.C.-V., J.M.-G., E.G.-R., I.H.-G. and S.G.-A.; resources, E.G.-R., I.H.-G. and S.G.-A.; writing—original draft preparation, J.P.-N.; writing—review and editing, P.M.I.-C., E.G.-R. and S.G.-A.; supervision, I.H.-G.; project administration, I.H.-G.; funding acquisition, I.H.-G. All authors have read and agreed to the published version of the manuscript.

Funding: This research was funded by the Castilla-La Mancha Regional Government project (POII-2014-008-P).

Acknowledgments: J.P.-N. thanks the European Social Fund and the University of Castilla-La Mancha for co-funding his predoc contract [2015/4062]. P.M.I.-C. thanks the European Social Fund and the Castilla-La Mancha Regional Government for co-funding his contract through the INCRECYT program. J.P.-N. is grateful to Carmen Verdejo for collaborating in the preparation of wine samples for phenolic compound analysis.

Conflicts of Interest: The authors declare no conflict of interest. The funders had no role in the design of the study; in the collection, analyses, or interpretation of data; in the writing of the manuscript, or in the decision to publish the results.

References

1. Lesschaeve, I. Sensory evaluation of wine and commercial realities: Review of current practices and perspectives. *Am. J. Enol. Vitic.* **2007**, *58*, 252–257.
2. Francis, L.; Osidacz, P.; Bramley, B.; King, E.; O'Brien, V.; Curtin, C.; Waters, E.; Jeffery, D.; Herderich, M.J.; Pretorius, I.S. Linking wine flavour components, sensory properties and consumer quality perceptions. *Aust. N. Z. Wine Ind. J.* **2010**, *25*, 18–23.
3. Soyollkham, B.; Valášek, P.; Fišera, M.; Fic, V.; Kubáň, V.; Hoza, I. Total polyphenolic compounds contents (TPC), total antioxidant activities (TAA) and HPLC determination of individual polyphenolic compounds in selected Moravian and Austrian wines. *Cent. Eur. J. Chem.* **2011**, *9*, 677–687. [CrossRef]
4. Santos-Buelga, C.; De Freitas, V. Influence of phenolics on wine organoleptic properties. In *Wine Chemistry and Biochemistry*; Moreno-Arrivas, M.V., Polo, M.C., Eds.; Springer: New York, NY, USA, 2009; pp. 529–570.
5. Petka, J.; Ferreira, V.; González-Viñas, M.Á.; Cacho, J. Sensory and chemical characterization of the aroma of a white wine made with Devín grapes. *J. Agric. Food Chem.* **2006**, *54*, 909–915. [CrossRef] [PubMed]
6. Liberatore, M.T.; Pati, S.; Del Nobile, M.A.; La Notte, E. Aroma quality improvement of Chardonnay white wine by fermentation and ageing in barrique on lees. *Food Res. Int.* **2010**, *43*, 996–1002. [CrossRef]
7. Di Maro, E.; Ercolini, D.; Coppola, S. Yeast dynamics during spontaneous wine fermentation of the Catalanesca grape. *Int. J. Food Microbiol.* **2007**, *117*, 201–210. [CrossRef] [PubMed]
8. Lukić, I.; Jedrejčić, N.; Ganić, K.K.; Staver, M.; Peršuric, D. Phenolic and aroma composition of white wines produced by prolonged maceration and maturation in wooden barrels. *Food Technol. Biotechnol.* **2015**, *53*, 407–418. [PubMed]
9. Biniari, K.; Gerogiannis, O.; Daskalakis, I.; Bouza, D.; Stavrakaki, M. Study of some qualitative and quantitative characters of the grapes of indigenous Greek grapevine varieties (*Vitis vinifera* L.) using HPLC and spectrophotometric analyses. *Not. Bot. Horti Agrobot. Cluj-Napoca* **2018**, *46*, 97–106.
10. Muccillo, L.; Gambuti, A.; Frusciante, L.; Iorizzo, M.; Moio, L.; Raieta, K.; Rinaldi, A.; Colantuoni, V.; Aversano, R. Biochemical features of native red wines and genetic diversity of the corresponding grape varieties from Campania region. *Food Chem.* **2014**, *143*, 506–513. [CrossRef] [PubMed]

11. Pérez-Navarro, J.; Izquierdo-Cañas, P.M.; Mena-Morales, A.; Martínez-Gascueña, J.; Chacón-Vozmediano, J.L.; García-Romero, E.; Hermosín-Gutiérrez, I.; Gómez-Alonso, S. Phenolic compounds profile of different berry parts from novel *Vitis vinifera* L. red grape genotypes and Tempranillo using HPLC-DAD-ESI-MS/MS: A varietal differentiation tool. *Food Chem.* **2019**, *295*, 350–360. [CrossRef]
12. Shiozaki, S.; Murakami, K. Lipids in the seeds of wild grapes native to Japan: *Vitis coignetiae* and *Vitis ficifolia* var. *ganebu*. *Sci. Hortic.* **2016**, *201*, 124–129. [CrossRef]
13. Mena, A.; Martínez, J.; Fernández-González, M. Recovery, identification and relationships by microsatellite analysis of ancient grapevine cultivars from Castilla-La Mancha: The largest wine growing region in the world. *Genet. Resour. Crop Evol.* **2014**, *61*, 625–637. [CrossRef]
14. Fernández-González, M.; Martínez-Gascueña, J.; Mena-Morales, A. Identification and relationships of grapevine cultivars authorized for cultivation in Castilla La Mancha (Spain). *Am. J. Enol. Vitic.* **2012**, *63*, 564–567. [CrossRef]
15. Pérez-Navarro, J.; Cazals, G.; Enjalbal, C.; Izquierdo-Cañas, P.M.; Gómez-Alonso, S.; Saucier, C. Flavanol glycoside content of grape seeds and skins of *Vitis vinifera* varieties grown in Castilla-La Mancha, Spain. *Molecules* **2019**, *24*, 4001. [CrossRef] [PubMed]
16. García-Carpintero, E.G.; Sánchez-Palomo, E.; Gallego, M.A.G.; González-Viñas, M.A. Volatile and sensory characterization of red wines from cv. Moravia Agria minority grape variety cultivated in La Mancha region over five consecutive vintages. *Food Res. Int.* **2011**, *44*, 1549–1560. [CrossRef]
17. Gómez Gallego, M.A.; Sánchez-Palomo, E.; Hermosín-Gutiérrez, I.; González Viñas, M.A. Polyphenolic composition of Spanish red wines made from Spanish *Vitis vinifera* L. red grape varieties in danger of extinction. *Eur. Food Res. Technol.* **2013**, *236*, 647–658. [CrossRef]
18. OIV. *Compendium of International Methods of Wine and Must Analysis*; International Organisation of Vine and Wine: Paris, France, 2014.
19. Ayala, F.; Echávarri, J.F.; Negueruela, A.I. A new simplified method for measuring the color of wines. I. Red and Rosé Wines. *Am. J. Enol. Vitic.* **1997**, *48*, 357–363.
20. Ough, C.S.; Amerine, M.A. *Methods Analysis of Musts and Wines*; John Wiley & Sons: New York, NY, USA, 1988.
21. Sarneckis, C.J.; Dambergs, R.G.; Jones, P.; Mercurio, M.; Herderich, M.J.; Smith, P.A. Quantification of condensed tannins by precipitation with methyl cellulose: Development and validation of an optimised tool for grape and wine analysis. *Aust. J. Grape Wine Res.* **2006**, *12*, 39–49. [CrossRef]
22. Lago-Vanzela, E.S.; Rebello, L.P.G.; Ramos, A.M.; Stringheta, P.C.; Da-Silva, R.; García-Romero, E.; Gómez-Alonso, S.; Hermosín-Gutiérrez, I. Chromatic characteristics and color-related phenolic composition of Brazilian young red wines made from the hybrid grape cultivar BRS Violeta ("BRS Rúbea"×"IAC 1398-21"). *Food Res. Int.* **2013**, *54*, 33–43. [CrossRef]
23. Pérez-Navarro, J.; Izquierdo-Cañas, P.M.; Mena-Morales, A.; Martínez-Gascueña, J.; Chacón-Vozmediano, J.L.; García-Romero, E.; Gómez-Alonso, S.; Hermosín-Gutiérrez, I. First chemical and sensory characterization of Moribel and Tinto Fragoso wines using HPLC-DAD-ESI-MS/MS, GC-MS, and Napping®techniques: Comparison with Tempranillo. *J. Sci. Food Agric.* **2019**, *99*, 2108–2123. [CrossRef]
24. Rebello, L.P.G.; Lago-Vanzela, E.S.; Barcia, M.T.; Ramos, A.M.; Stringheta, P.C.; Da-Silva, R.; Castillo-Muñoz, N.; Gómez-Alonso, S.; Hermosín-Gutiérrez, I. Phenolic composition of the berry parts of hybrid grape cultivar BRS Violeta (BRS Rubea×IAC 1398-21) using HPLC-DAD-ESI-MS/MS. *Food Res. Int.* **2013**, *54*, 354–366. [CrossRef]
25. Ibarz, M.J.; Ferreira, V.; Hernández-Orte, P.; Loscos, N.; Cacho, J. Optimization and evaluation of a procedure for the gas chromatographic-mass spectrometric analysis of the aromas generated by fast acid hydrolysis of flavor precursors extracted from grapes. *J. Chromatogr. A* **2006**, *1116*, 217–229. [CrossRef] [PubMed]
26. Marchante, L.; Loarce, L.; Izquierdo-Cañas, P.M.; Alañón, M.E.; García-Romero, E.; Pérez-Coello, M.S.; Díaz-Maroto, M.C. Natural extracts from grape seed and stem by-products in combination with colloidal silver as alternative preservatives to SO_2 for white wines: Effects on chemical composition and sensorial properties. *Food Res. Int.* **2019**, *125*, 108594. [CrossRef] [PubMed]
27. Izquiedo Cañas, P.M.; Mena Morales, A.; Heras Manso, J.M.; García Romero, E.; González Viñas, M.A.; Sánchez Palomo, E. Chemical and sensory characterization of the aroma of "Chardonnay" musts fermented with different nitrogen sources. *Cienc. Tec. Vitivinic.* **2018**, *33*, 116–124. [CrossRef]
28. *ISO-8586-1 Sensory Analysis—General Guidance for the Selection, Training, and Monitoring of Assessors. Part 1: Selected Assessors*; International Standardization Office: Geneva, Switzerland, 1993.

29. *ISO-11035 Sensory Analysis—Identification and Selection of Descriptors for Establishing a Sensory Profile by Multidimensional Approach*; International Standardization Office: Geneva, Switzerland, 1994.
30. *ISO-8589 Sensory Analysis—General Guidance for the Design of Test Roomle*; International Standardization Office: Geneva, Switzerland, 2007.
31. Pagès, J. Collection and analysis of perceived product inter-distances using multiple factor analysis: Application to the study of 10 white wines from the Loire Valley. *Food Qual. Prefer.* **2005**, *16*, 642–649. [CrossRef]
32. Sánchez-Palomo, E.; Alonso-Villegas, R.; Delgado, J.A.; González-Viñas, M.A. Improvement of Verdejo white wines by contact with oak chips at different winemaking stages. *LWT Food Sci. Technol.* **2017**, *79*, 111–118. [CrossRef]
33. Hermosín-Gutiérrez, I.; Castillo-Muñoz, N.; Gómez-Alonso, S.; García-Romero, E. Flavonol profiles for grape and wine authentication. In *Progress in Authentication of Food and Wine*; ACS Symposium Series; Ebeler, S.E., Takeoka, G.R., Winterhalter, P., Eds.; American Chemical Society: Washington, DC, USA, 2011; pp. 113–129.
34. Cejudo-Bastante, M.J.; Pérez-Coello, M.S.; Pérez-Juan, P.M.; Hermosín-Gutiérrez, I. Effects of hyper-oxygenation and storage of Macabeo and Airén white wines on their phenolic and volatile composition. *Eur. Food Res. Technol.* **2012**, *234*, 87–99. [CrossRef]
35. Castillo-Muñoz, N.; Gómez-Alonso, S.; García-Romero, E.; Hermosín-Gutiérrez, I. Flavonol profiles of *Vitis vinifera* red grapes and their single-cultivar wines. *J. Agric. Food Chem.* **2007**, *55*, 992–1002. [CrossRef]
36. Schwarz, M.; Picazo-Bacete, J.J.; Winterhalter, P.; Hermosín-Gutiérrez, I. Effect of copigments and grape cultivar on the color of red wines fermented after the addition of copigments. *J. Agric. Food Chem.* **2005**, *53*, 8372–8381. [CrossRef]
37. Kennedy, J.A.; Saucier, C.; Glories, Y. Grape and wine phenolics: History and perspective. *Am. J. Enol. Vitic.* **2006**, *57*, 239–248.
38. Cheynier, V.F.; Trousdale, E.K.; Singleton, V.L.; Salgues, M.J.; Wylde, R. Characterization of 2-5-glutathionylcaftaric acid and its hydrolysis in relation to grape wines. *J. Agric. Food Chem.* **1986**, *34*, 217–221. [CrossRef]
39. Cejudo-Bastante, M.J.; Hermosín-Gutiérrez, I.; Castro-Vázquez, L.I.; Pérez-Coello, M.S. Hyperoxygenation and bottle storage of Chardonnay white wines: Effects on color-related phenolics, volatile composition, and sensory characteristics. *J. Agric. Food Chem.* **2011**, *59*, 4171–4182. [CrossRef] [PubMed]
40. Romero-Pérez, A.I.; Lamuela-Raventós, R.M.; Waterhouse, A.L.; de la Torre-Boronat, M. Levels of *cis*- and *trans*-resveratrol and their glucosides in white and rosé *Vitis vinifera* wines from Spain. *J. Agric. Food Chem* **1996**, 2124–2128. [CrossRef]
41. Ferreira, V.; Fernández, P.; Peña, C.; Escudero, A.; Cacho, J.F. Investigation on the role played by fermentation esters in the aroma of young Spanish wines by multivariate analysis. *J. Sci. Food Agric.* **1995**, *67*, 381–392. [CrossRef]
42. Guth, H. Identification of character impact odorants of different white wine varieties. *J. Agric. Food Chem.* **1997**, *45*, 3022–3026. [CrossRef]
43. Sánchez Palomo, E.; Díaz-Maroto Hidalgo, M.C.; González-Viñas, M.Á.; Pérez-Coello, M.S. Aroma enhancement in wines from different grape varieties using exogenous glycosidases. *Food Chem.* **2005**, *92*, 627–635. [CrossRef]
44. Ferreira, V.; Lopez, R.; Aznar, M. Olfactometry and aroma extract dilution analysis of wines, molecular methods of plant analysis. In *Analysis of Taste and Aroma*; Jackson, J., Linskens, H., Eds.; Springer: Berlin/Heidelberg, Germany, 2002; pp. 88–122.
45. Ferreira, V.; López, R.; Cacho, J.F. Quantitative determination of the odoorants of young red wines from different grape varieties. An assessment of their sensory role. *J. Sci. Food Agric.* **2000**, *80*, 1659–1667. [CrossRef]
46. Etiévant, P.X. Wine. In *Volatile Compounds of Food and Beverages*; Maarse, H., Ed.; Marcel Dekker: New York, NY, USA, 1991; pp. 483–546.
47. Gómez-Míguez, M.J.; Cacho, J.F.; Ferreira, V.; Vicario, I.M.; Heredia, F.J. Volatile components of Zalema white wines. *Food Chem.* **2007**, *100*, 1464–1473. [CrossRef]
48. Rapp, A.; Versini, G. Influence of nitrogen compounds in grapes on aroma compounds of wines. *Dev. Food Sci.* **1995**, *37*, 1659–1694.

49. Sánchez-Palomo, E.; Díaz-Maroto, M.C.; Viñas, M.A.G.; Soriano-Pérez, A.; Pérez-Coello, M.S. Aroma profile of wines from Albillo and Muscat grape varieties at different stages of ripening. *Food Control* **2007**, *18*, 398–403. [CrossRef]
50. Perestrelo, R.; Fernandes, A.; Albuquerque, F.F.; Marques, J.C.; Câmara, J.S. Analytical characterization of the aroma of Tinta Negra Mole red wine: Identification of the main odorants compounds. *Anal. Chim. Acta* **2006**, *563*, 154–164. [CrossRef]
51. Peinado, R.A.; Moreno, J.; Medina, M.; Mauricio, J.C. Changes in volatile compounds and aromatic series in sherry wine with high gluconic acid levels subjected to aging by submerged flor yeast cultures. *Biotechnol. Lett.* **2004**, *26*, 757–762. [CrossRef] [PubMed]
52. Escudero, A.; Gogorza, B.; Melús, M.A.; Ortín, N.; Cacho, J.; Ferreira, V. Characterization of the aroma of a wine from Maccabeo. Key role played by compounds with low odor activity values. *J. Agric. Food Chem.* **2004**, *52*, 3516–3524. [CrossRef]
53. Escudero, A.; Campo, E.; Fariña, L.; Cacho, J.; Ferreira, V. Analytical characterization of the aroma of five premium red wines. Insights into the role of odor families and the concept of fruitiness of wines. *J. Agric. Food Chem.* **2007**, *55*, 4501–4510. [CrossRef] [PubMed]

Review

The Flavor Chemistry of Fortified Wines—A Comprehensive Approach

Teresa Abreu [1], Rosa Perestrelo [1], Matteo Bordiga [2], Monica Locatelli [2], Jean Daniel Coïsson [2] and José S. Câmara [1,3,*]

[1] CQM–Centro de Química da Madeira, Universidade da Madeira, Campus Universitário da Penteada, 9020-105 Funchal, Portugal; teresa.abreu@staff.uma.pt (T.A.); rmp@staff.uma.pt (R.P.)

[2] Dipartimento di Scienze del Farmaco, Università degli Studi del Piemonte Orientale "A. Avogadro", Largo Donegani 2, 28100 Novara, Italy; matteo.bordiga@uniupo.it (M.B.); monica.locatelli@uniupo.it (M.L.); jeandaniel.coisson@uniupo.it (J.D.C.)

[3] Departamento de Química, Faculdade de Ciências Exatas e Engenharia, Campus da Penteada, Universidade da Madeira, 9020-105 Funchal, Portugal

* Correspondence: jsc@staff.uma.pt; Tel.: +351-(291)-705112

Abstract: For centuries, wine has had a fundamental role in the culture and habits of different civilizations. Amongst numerous wine types that involve specific winemaking processes, fortified wines possess an added value and are greatly honored worldwide. This review comprises the description of the most important characteristics of the main worldwide fortified wines—Madeira, Port, Sherry, Muscat, and Vermouth—structured in three parts. The first part briefly describes the chemistry of wine flavor, the origin of typical aroma (primary, secondary and tertiary), and the influencing parameters during the winemaking process. The second part describes some specificities of worldwide fortified wine, highlighting the volatile composition with particular emphasis on aroma compounds. The third part reports the volatile composition of the most important fortified wines, including the principal characteristics, vinification process, the evolution of volatile organic compounds (VOCs) during the aging processes, and the most important odor descriptors. Given the worldwide popularity and the economic relevance of fortified wines, much research should be done to better understand accurately the reactions and mechanisms that occur in different stages of winemaking, mainly during the oxidative and thermal aging.

Keywords: flavor; aroma origin; fortified wines; enology; vinification process; aging; odor descriptors; Madeira wines

Citation: Abreu, T.; Perestrelo, R.; Bordiga, M.; Locatelli, M.; Daniel Coïsson, J.; Câmara, J.S. The Flavor Chemistry of Fortified Wines—A Comprehensive Approach. *Foods* **2021**, *10*, 1239. https://doi.org/

Academic Editor: Panagiotis Kandylis

Received: 17 March 2021
Accepted: 26 May 2021
Published: 29 May 2021

Publisher's Note: MDPI stays neutral with regard to jurisdictional claims in published maps and institutional affiliations.

1. A Short Introduction about Wine Flavor Chemistry

Chemically, wine is a fascinating and very complex matrix constituted by several hundreds of chemical compounds/groups—water, ethanol, glycerol, organic acids, carbohydrates, and, to a minor extent, terpenoids, pyrazines, higher alcohols, ethyl esters, fatty acids, nitrogenous compounds, sulphur compounds, furanic compounds, among others. These chemical groups were found in a broad range of concentrations (from a few mg/L to ng/L) and presenting different polarities and volatilities [1–4]. Table 1 shows the main volatile organic compounds (VOCs) found in wines. Some VOCs are responsible for the singular aromatic characteristics of some wine types. It should be mentioned that the most pleasant wines hardly have precise and simple to identify aromas; rather, they have complex aromatic aroma descriptors in which some fruit and floral perception is crucial, along with other spicy, woody, and/or toast notes. It should also be pointed out that the lack of aromatic faults or variations is also a constant and vital factor of quality [1,5].

Table 1. Some important volatile organic compounds (VOCs) identified in fortified wines, chemical structure, aroma descriptors, and odor threshold (OT) [4–11].

Chemical Groups	VOCs	Chemical Structure	Aroma Descriptor	OT [1] (μg/L)
Terpenoids	α-Terpeniol	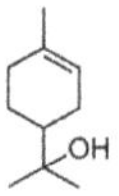	Floral, linden	250
	Linalool	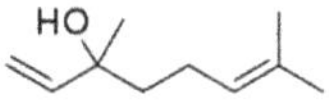	Rose	25
	Geraniol	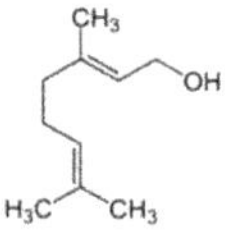	Citrus, floral	20
	Citronellol	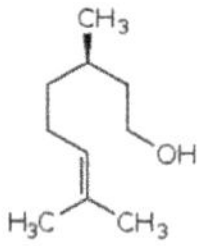	Citrus, floral, sweet	100
	Rotundone	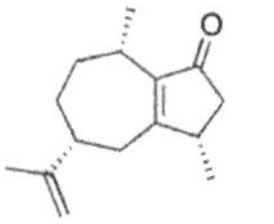	Pepper, spicy	0.016
C13 Norisoprenoids	β-Damascenone	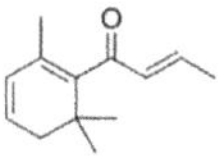	Boiled apple, sweet	0.05
	2,6,6-Trimethylcyclohex-2-ene-1,4-dione	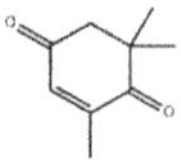	Musty, citrus, sweet honey	25
	TDN [2]	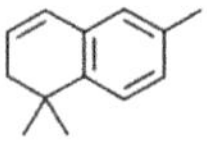	Floral, fruit	4
	Vitispirane	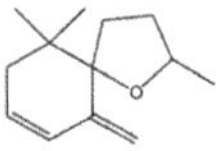	Floral, spice, wood	800
C6 compounds	1-Hexanol	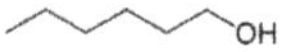	Herbaceous	8000
	(Z)-3-Hexen-1-ol	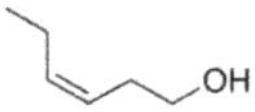	Green, bitter	400
	1-Hexanal		Green	97

Table 1. *Cont.*

Chemical Groups	VOCs	Chemical Structure	Aroma Descriptor	OT [1] (μg/L)
Higher alcohols	3-Methylbutanol	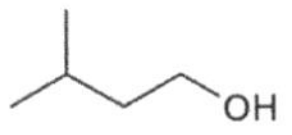 OH	Fusel, sour	30,000
	2-Phenylethanol	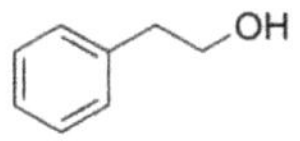 OH	Rose, honey	140,000
	Benzyl alcohol	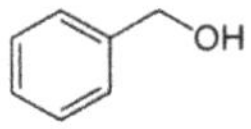 OH	Blackberry	200,000
Ethyl esters	Ethyl acetate	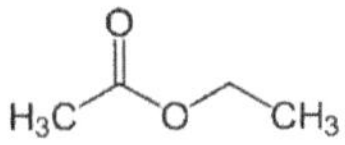 O, H_3C, O, CH_3	Solvent, nail polish, fruity	12,000
	Ethyl hexanoate	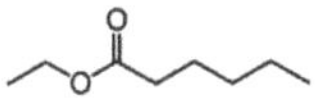 O, O	Green apple	14
	Ethyl octanoate	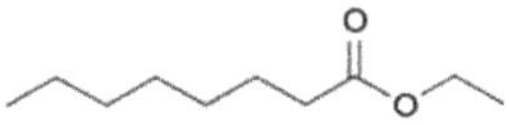 O, O	Sweet, flower	2
	Ethyl decanoate	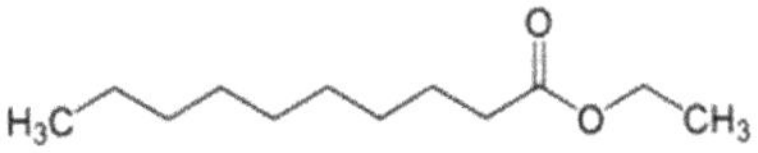 O, H_3C, O, CH_3	Brandy, grape	200
	Ethyl lactate	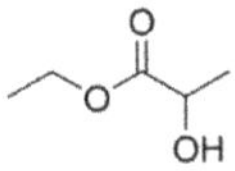 O, O, OH	Butter	150,000
	Diethyl succinate	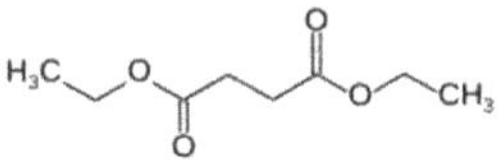 H_3C, O, O, O, O, CH_3	Melon	500,000
Acetates	Isoamyl acetate	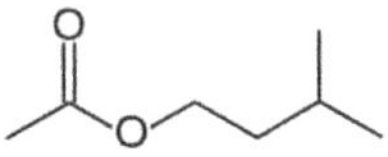 O, O	Banana, sweet	160
	2-Phenylethyl acetate	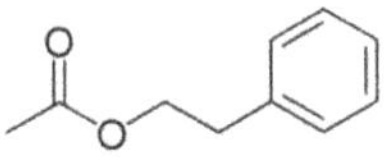 O, O	Rose, flower	1800

Table 1. *Cont.*

Chemical Groups	VOCs	Chemical Structure	Aroma Descriptor	OT [1] (μg/L)
Acids	Acetic acid	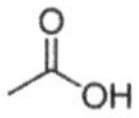 O OH	Vinegar, sour	200,000
	Hexanoic acid	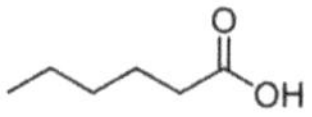 O OH	Cheese, fatty	3000
	Octanoic acid	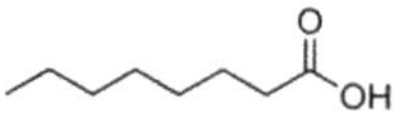 O OH	Fatty, rancid	10,000
	Decanoic acid	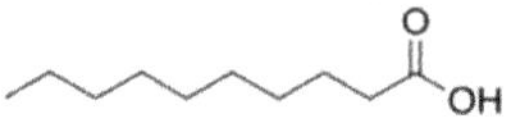 O OH	Fatty, rancid	15,000
Carbonyl compounds	2-Phenylacetaldehyde	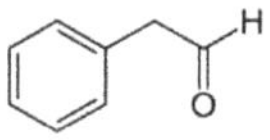 H O	Floral	1
	Diacetyl	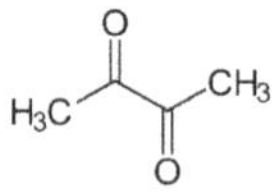 O H_3C CH_3 O	Buttery	100
	Benzaldehyde	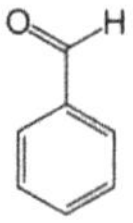 O H	Almond	2000
Furanic compounds	2-Furfural	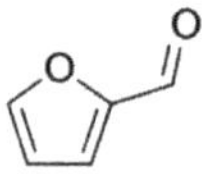 O O	Wood, nut	14,100
	5-Methyl-2-furfural	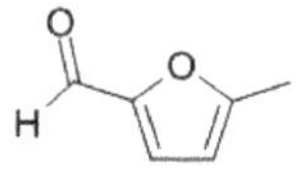 O H O	Caramel	20,000
	HMF [3]	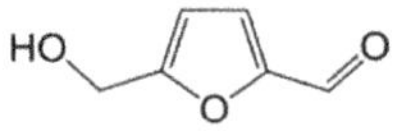 HO O O	Almond, nut	10,000
Volatile phenols	4-Vinyl-guaiacol	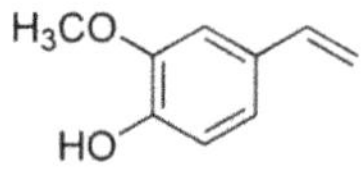 H_3CO HO	Smoke, phenolic	40
	Methyl vanillate	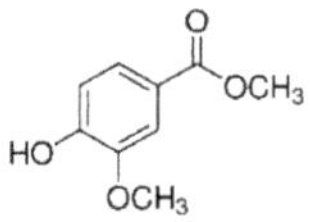 O OCH_3 HO OCH_3	Vanillin	3000
	4-Vinyl-phenol	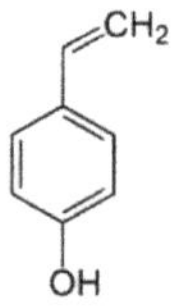 CH_2 OH	Spicy, pharmaceutical	180

Table 1. *Cont.*

Chemical Groups	VOCs	Chemical Structure	Aroma Descriptor	OT [1] (μg/L)
Lactones	Sotolon	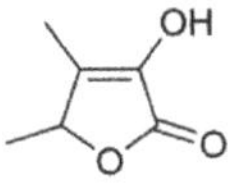	Wood, nut, toast	19
	γ-Decalactone	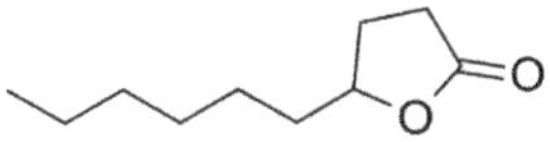	Peach	88
	Whisky lactone	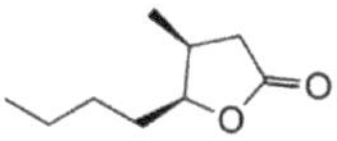	Caramel, nut, toast	67
Pyrazines	IBMP [4]	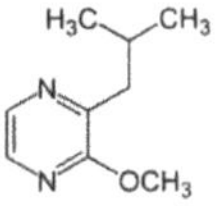	Leafy	0.016
	SBMP [5]	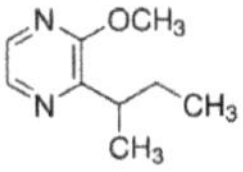	Green, pepper	0.002
	IPMP [6]	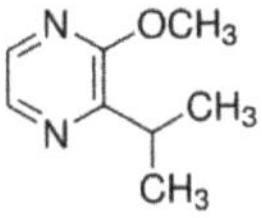	Leafy	0.002
Sulphur compounds	3-Mercaptohexyl acetate	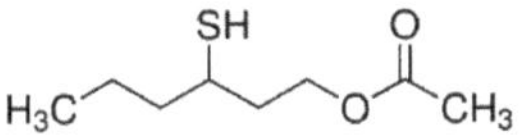	Passion fruit, box tree	0.0042
	4-(Methylthio)-4-methyl-2-pentanone	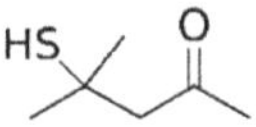	Box tree, tropical fruit	0.0008
	3-Mercaptohexanol	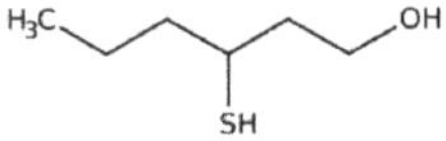	Passion fruit, grapefruit	0.06

[1] OT—Odor threshold was determined using a synthetic wine model with an ethanol content ranging from 10 to 12% of ethanol; [2] TDN: 1,2-dihydro-1,1,6-trimethylnaphthalene; [3] HMF-5-hydroxymethyl-2-furfural; [4] IBMP: 3-isobutyl-2-methoxypyrazine; [5] SBMP: 3-sec-butyl-2-methoxypyrazine; and [6] IPMP: 3-isopropryl-2-methoxypyrazine.

The primary and secondary metabolites, which have been identified in grapes, musts, and wines, are synthesized throughout several pathways occurring from the vineyard to the consumer, including (i) biosynthesis in grapes; (ii) yeast metabolism during fermentative process; and (iii) several enzymatic and chemical reactions occurring during aging. Some of these metabolites actually contribute to the sensory perception of wine flavor and its specific organoleptic characteristics, being responsible for the wine aroma quality [1–3].

The final quality of wine depends on several factors and parameters, namely the grape varieties, geographical region, terroir and climatological conditions, the vinification process including fermentation conditions (must composition, dominant yeasts, pH, temperature), and aging. Terpenoids and their derivatives constitute significant markers of grape quality, contributing floral notes to the wine flavor and aroma when present in amounts higher than its odor threshold (OT) [12]. Nevertheless, several of the sensory properties that are commonly used to evaluate the quality of wine, including those that are deemed characteristic of the grape variety, are not commonly perceived in the grapes. They appear, mainly, through different chemical and biochemical pathways that occur during the winemaking process. The main key odorants and off-flavors, their impact on wine authenticity, and their evolution through aging, has been the focus of deep research and scientific interest. In addition, the interactions of odorants with other non-volatile wine compounds, and their effect on aroma quality, is also emerging as a potential field of curiosity in the scientific community. A deeper and comprehensive understanding of the biochemistry of grape-juice fermentation, and the chemistry of wine aging, is of utmost importance to help the wine industry by reinforcing the empirical knowledge of the traditional winemaker [13]. Different wine metabolomes can be used as useful tools in the research of key chemical components of a particular wine, allowing us to differentiate it from other wines [14–17]. In addition, they make it possible to connect chemical composition with the sensory properties, either by detecting influence VOCs or clarifying matrix effects.

The low concentration of key aroma compounds, such as esters, terpenoids, pyrazines, and thiols, in addition to the low odor threshold (OT) of many important contributors to the aroma, and the influence of alcohol matrix in the odor perception, make the accurate flavor definition of wine extremely challenging [5]. With the advance of analytical instrumentation, particularly the greater accessibility of gas chromatography-mass spectrometry (GC-MS), new insights about the flavor and its precursors in alcoholic beverages (e.g., wine) have been achieved [5,18–20].

To assess the impact of individual VOCs on the global flavor of wine, beverages, food, and other food-related samples, aroma extract dilution analysis (AEDA) [21–23] is one of the most appropriate and commonly-used procedures. However, the use of solvent-based extractions presents some drawbacks related to the loss of highly VOCs during the extraction and/or concentration of organic extract procedure, typically carried out under a nitrogen stream, in addition to the use of harmful organic solvents to both the operator and the environment. To overcome these shortcomings, emerging strategies of flavor dilution analysis, headspace solid-phase microextraction (HS-SPME) [1,24–26], and stir bar sorptive extraction (SBSE) [27–29] have been used with increasing frequency.

The Wine Flavor: Origin and Influencing Parameters

The flavor of wine is influenced by parameters including environmental factors (e.g., soil, climate), vineyard location, pre-fermentation biochemical phenomena (e.g., oxidations, hydrolysis), fermentation conditions (e.g., pH, temperature, microflora), vinification techniques, several post-fermentation processes (e.g., clarification, fining, filtration) to which the wine is submitted, and by the storage (humidity and temperature) and aging conditions (Figure 1) [17–21].

Based on their origin, the wine flavor can be categorized into (i) primary or varietal aroma, for compounds biosynthesized in grape; (ii) pre-fermentative aroma, formed during the processing of the grape harvest and subsequent operations; (iii) fermentative or secondary aroma synthesized during alcoholic fermentation by yeasts; and (iv) tertiary or aging aroma formed from enzymatic and physicochemical reactions such as oxidation and reduction, which occur during the conservation and wine aging [1,22].

The primary aroma is related to a complex set of compounds that can appear in free form, which contribute directly to odor (VOCs, odorants) and/or as odorless nonvolatile glycosides (glycosidically bound compounds) [1]. The acid and/or enzymatic hydrolysis of the glycosidically bound compounds may give raise to odorant aglycones during the winemaking process and aging, by the action of endogenous and/or exogenous glycosidase enzymes, or be caused by the moderate acidic conditions of grape juice and wine [23]. The primary aroma is biosynthesized during grape ripening. Their chemical nature and concentration depend on several parameters including soil, climate, phytotechnology and physiology of the vineyard, grape health status, and degree of ripeness of the grape. These types of aroma compounds contribute to the quality and aromatic typicity of young wines. Most of them are terpenoids (characteristic of Muscatel grapes and wines), C13-

norisoprenoids (abundant in Chardonnay), and methoxypyrazines (characteristic of the Cabernet grapes and wines) [24,25]. In addition, some thiols have been identified as important contributors to the aroma of numerous grape varieties, namely Cabernet-Sauvignon, Sauvignon Blanc, and Merlot [26].

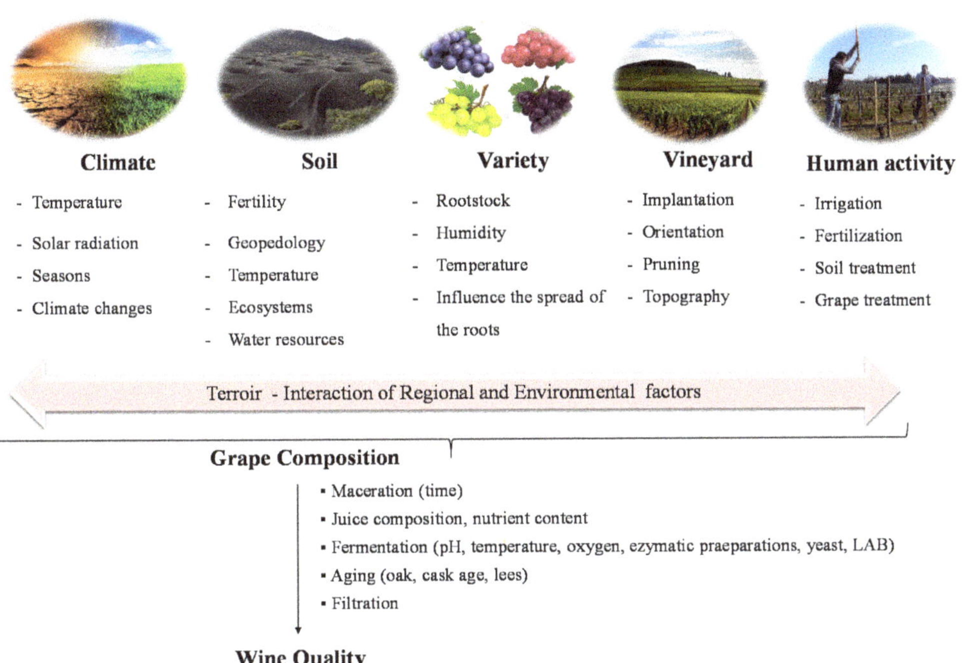

Figure 1. Most important influencing parameters on wine quality and aroma.

Pre-fermentative aroma, namely C6 aldehydes and C6 alcohols, are biosynthesized in the period between harvest and fermentation, due to the action of endogenous enzymes, as a result of the differentiated technological operations to which the grapes are submitted, namely harvesting, transport, crushing, and pressing. These operations allow the grapes enzymatic system interact with the membraneer lipids, releasing the polyunsaturated fatty acids—linoleic and linolenic acids [1,17,27], which, via oxidation, originate the C6 compounds. The concentration of C6 alcohols depends on grape variety, ripeness stage, treatment prior to fermentation, and temperature/duration of contact with skins [27]. The ratio (*E*)-3-hexenol/(*Z*)-3-hexenol may be useful for wine differentiation according to grape variety [17].

Fermentative aroma is produced by the action of yeast during alcoholic fermentation and by lactic bacteria if malolactic fermentation takes place. In the course of this process, the wines' chemical and aromatic complexity increases significantly as a product of chemical transformations of the grape and must constituents, and as a result of the metabolism of the fermentative yeasts or bacteria (malolactic fermentation). Generally, the fermentation process is started and catalyzed by numerous indigenous yeast strains, namely *Saccharomyces*, *Kloeckera*, *Candida*, *Hansenula*, *Hanseniospora*, *Pichia*, and *Zigossacharimyce*. The yeast development during fermentation depends on must pH, fermentation temperature, nitrogen level, sulphur dioxide, and sugar content [19,28]. Besides ethanol and glycerol, several secondary metabolites are formed during the fermentation process that, while modest in quantity, can contribute significantly to the overall aroma of the wine. These include higher alcohols, fatty acids, and ethyl esters from volatile and fix acids, and, to a lesser extent, carbonyl compounds (e.g., aldehydes and ketones), lactones, and volatile phenols [1]. Malolactic fermentation is a secondary fermentation which takes place after alcoholic fermentation, and was generally performed in wines with enhanced acidity. The malolactic fermentation is conducted by malolactic bacteria (MLB), most often strains of *Oenococcus oeni*, and involves the decarboxylation of L-malic acid into L-lactic acid [29].

The compounds related to tertiary aroma arise during the wine maturation/conservation period. Throughout conservation, the volatile composition of the wine undergoes significant changes responsible for the evolution of the aroma, progressively losing the fruity and floral descriptors typical of young wines, and evolving to a more complex *bouquet* with caramel, dried fruit, spicy, toasty, and wood aroma notes. Some wines acquire their aromatic typicity after several years of aging, while others cannot stand long storage periods [30]. The excellence of the aging bouquet varies on wine origin (e.g., vineyard soil, microclimate), vintage, and diffusion of different compounds from oak to wine [21,31]. This diffusion process depends on the oak properties (e.g., geographic origin, oak species, seasoning of the staves, toasting, and cask age) [31,32]. The most common woods used during wine aging are obtained from different oak species such as *Quercus alba*, *Q. robur*, or *Q. petrea*, but also from other species known to contain high contents of ellagitannins, namely *Acacia*, *Castanea*, or *Prunus* [33]. The furanic compounds (e.g., 2-furfural, 5-hydroxymethylfurfural), lactones (e.g., sotolon, whiskylactone), and acetals (e.g., dioxane isomers, dioxolane isomers) are the chemical groups significantly associated to the evolution of wine ageing bouquet [20,34–36].

2. Fortified Wines

Fortified wines are characterized by a high alcohol content (between 15–22%, *v/v*), resulting from the addition of distilled spirits, usually a neutral grape spirit, and produced under oxidative conditions which determine the fortified wines' typical flavor and aroma profile. This type of wine can be produced using both fermented (partially and/or totally) and unfermented grape must, according to EU regulation No. 479/2008. Following the intrinsic characteristic of the wines, they are usually classified as dessert, fortified, and generoso. Regarding the former, since these are generally consumed at the end of the meal, the definition of dessert wines reflects their widespread use. Regarding the second, due to their high alcohol content, these wines acquire the definition of fortified. Regarding the latter, the term generoso is relative to a significant alcohol and sugar content present in wines. Traditionally, fortified wines originate from Europe, however their production is spreading worldwide [37]. Although this spread is growing, some European countries remain the main producers [38]. Sherry, Port, Madeira, and Marsala are the most common types of fortified wines, whereas Vermouth (Italy), Moscatel de Setúbal (Portugal), and Commandaria (Cyprus), are produced in lower amounts.

2.1. Sherry Wine

Sherry (15–22% alcohol by volume) is produced in the province of Cádiz, Andalusia (south west Spain) by using white grapes cultivated in the Sherry triangle (Sanlúcar de Barrameda, Jerez de la Frontera and El Puerto de Santa Maria) [37,39,40]. The white grapes used are primarily from Palomino (both *Jerez* and *Fino*) however, Moscatel and Pedro Ximénez grapes are also utilized. Figure 2 reports the main winemaking steps for the production of Sherry wines. In the same way as those of still wines, the various winemaking steps follow this order: grape crushing, pressing, fermentation, racking, and the fortification step, followed by aging [40]. Once the base wine is ready, the wine is classified by professional tasters in order to define the most appropriate type of aging for each sample. Once classified, in order to prevent spoilage (e.g., acetic acid bacteria), the fortification step can begin. The addition of alcohol is defined according to the diverse types of aging the wines are going to receive. In detail, with an ethanol content set up to 15% (as in the case of *Fino* wine), the yeast cells are able to tolerate this percentage, thus maintaining the flor velum (biological aging). Conversely, if set to about 18% (as in the case of *Oloroso* wine), oxidative aging occurs, since the growth of flor yeasts is prevented [41].

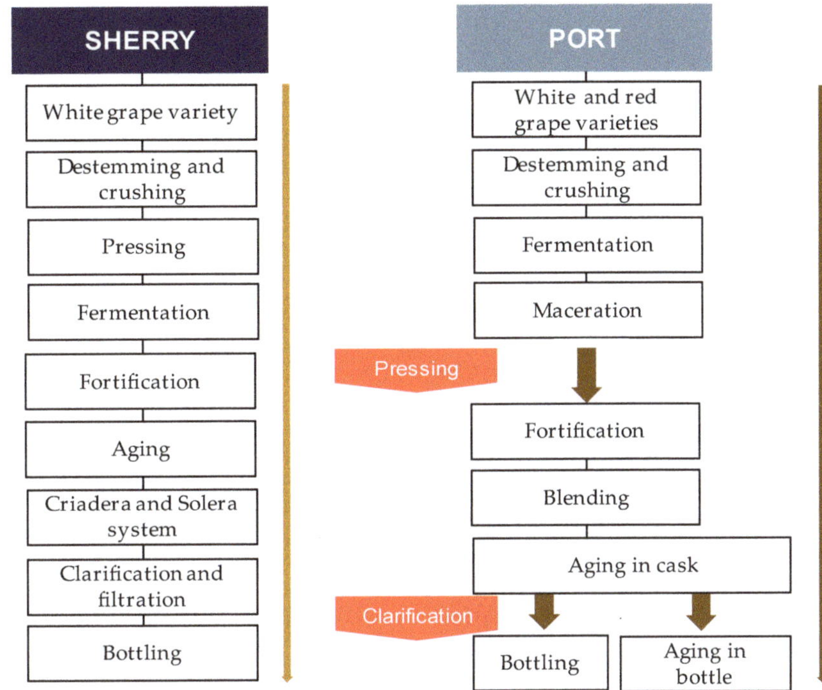

Figure 2. Sherry and Port's winemaking processes.

Biological aging involves a static phase ("añadas") during which the wine is kept in a butt for a variable number of years, followed by a dynamic one ("criaderas-solera"), consisting of a series of oak barrels in process of maturation. Nowadays, the solera system, established in the middle of the 19th century, is generally linked with modern Sherry production [42], however it is also utilized for other productions (e.g., Port, Madeira, brandy, and vinegar). In the initial stage, when the wine is young, about 15% (v/v) is defined as "sobretablas". The middle stages are called "criaderas", and the final one (also called "solera") contains the oldest wine (Figure 3).

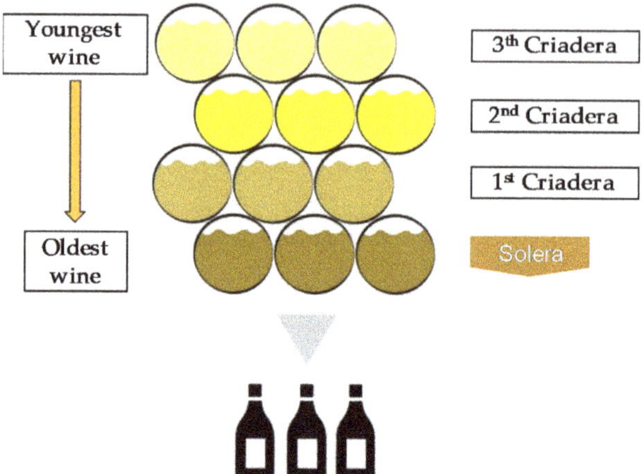

Figure 3. Sherry and Port's winemaking processes.

The *saca* process consists in removing a portion of wine from the solera by replacing it with an equal portion of younger wine from the first criadera. Generally, the volume of wine transferred ranges between 10–20% of the barrels capacity (about 500 L), even if, by law, up to 30% is allowed. The average wine age (from a minimum of 3 to 30 years) is determined by calculating the total system volume to annual volume removed ratio [41]. The solera system can be achieved through two different aging conditions: oxidative and biological [41]. Clarification represents the last phase before bottling. *Fino*, *Amontillado*, and *Oloroso* are the main styles of Sherry wines, characterized by certain peculiarities.

Fino is obtained by a gentle pressing of Palomino grapes; the wines appear fine and pale. The maturation period lasts between 3 and 8 years. Generally, the wines are left dry, characterized by a final alcoholic content ranging between 15.5–17% alcohol by volume [42,43]. *Fino* wines show a pale to light gold color, and they are characterized by a pungent apple flavor and peculiar notes such as fruity, floral, and cheesy [41,44].

Generally, *Oloroso*, produced using Palomino grapes obtained in hot climates, is darker, with a greater structure containing more phenols [42]. Among all Sherry styles, *Oloroso* is the most oxidized. The wines follow an oxidative aging process as their base wines are fortified up to 18% alcohol by volume. Its color varies from mahogany to amber, showing a complex bouquet characterized by distinct notes such as nutty, spicy, pungent smoke, tobacco, and leather [42,44]. Normally, the sweetest Sherries (about 20% alcohol by volume) are prepared starting from *Oloroso* wines by blending these with concentrated rectified must or naturally sweet wines [43].

Amontillado is characterized by a two-step process: (i) maturation under the flor velum (typical of *Finos*); (ii) oxidation period in the absence of flor [41]. From the organoleptic point of view, this wine is darker than *Fino* (ranging from amber to pale topaz). Thanks to the complexity of its bouquet, and due to the aging processes (*Fino* and *Oloroso*), *Amontillado* is the most valued Sherry style [45].

2.2. Port Wine

Port wine (15–22% alcohol by volume) is produced in the Douro Protected Designation of Origin (PDO) region (created and regulated in 1756). Lower Corgo, Upper Corgo, and Upper Douro represent the three distinct sub regions of the latter [46]. Apart from the Douro PDO region, other countries (e.g., Australia, South Africa, and the United States) produce Port-style wines by using techniques similar to those used in Portugal [47]. Portuguese varieties are commonly used in South Africa [46]. Among the many authorized grape cultivars, only six red ones are generally used to produce this kind of wine, namely Touriga

Francesa, Tinto Cão, Touriga Nacional, Tinta Barocca, and Tinta Roriz [37,48]. Thanks to easy adaptation to the region's soil characteristics, Touriga Franca represents half of the Douro PDO area. This variety shows a characteristic ruby color connected with complex notes such as cherry, raspberry, and blackberry [48]. Figure 2 schematized the main steps of Port winemaking. The fermentation temperatures of this type of wine range between 26–28 °C [38]. The singularity in Port production relies on the *aguardente* (wine spirit of 76–78% alcohol by volume; local definition relating to alcohol used in the fortification step). Once the amount of remaining sugars reaches the desired degree of sweetness (usually within 2–3 days), and having been evaluated by a panel for classification, the wines can begin the proper fortification phase [48]. According to the method of aging, wines can be assigned to two main categories: Ruby and Tawny style. The latter ages in smaller vessels (about 600 L; wooden) and the former in large ones (3–5 years followed by bottling). Tawny-style Ports show nutty, spicy, and woody notes, and specific brown and yellow hues. Ruby-style Ports show a red color, full-bodied structure, and fruity notes [38,49].

2.3. Madeira Wine

Boal, Malvasia, Sercial, and Verdelho (white grape varieties), known as noble varieties, and Tinta Negra (TN, red grape variety) were the main *V. vinifera* L. grapes used in the production of Madeira wines. Some volatile compounds identified in the pulp and skin of these grape varieties are summarized in Figure 4. Tinta negra is the most predominant, constituting more than 80% of the vineyards, as it is very resistant to diseases (e.g., oidium, phylloxera) [50]. There are still other minor recommended (e.g., Bastardo, Verdelho Tinto) and authorized (e.g., Complexa, Deliciosa, Listrão) varieties to produce Madeira wine [50]. Terrantez and Bastardo are used rarely, even though they produce wines of distinctive quality.

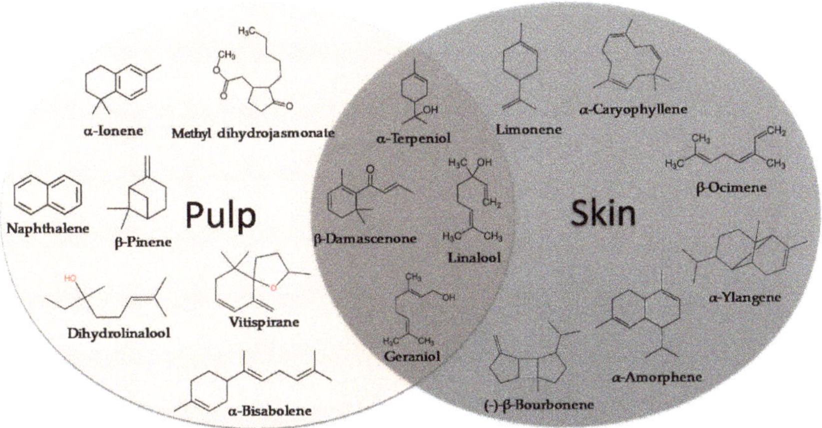

Figure 4. Some varietal compounds identified in pulp and skin of white *V. vinifera* L. grapes used in the production of Madeira wine [12,51].

Grape volatile composition [51] is influenced by the presence of hundreds of VOCs involving different chemical groups, among which terpenoids and C13 norisoprenoids are the most representative for most of *V. vinifera* L. varieties [51]. These chemical groups were found to be dispersed between the pulp and skin of grapes, being in higher concentration in the skin than in the pulp [12]. Perestrelo et al. [12,51] recognized the varietal pattern of *V. vinifera* L. grape varieties from different geographical origins, and the results demonstrated that the amounts and number of VOCs in the skin are remarkably higher than those detected in the pulp. Moreover, different varietal volatile profiles were observed for the investigated grapes (Figure 4), independently of grape fraction. This difference could be explained by climatic factors (e.g., altitude, temperature, or humidity). Boal showed the highest levels of sesquiterpenic compounds, whereas Malvasia and Malvasia Cândida showed the highest level of monoterpenic compounds.

According to Perestrelo et al. [52], the grape variety appears to have a remarkable influence in the sensorial characteristics of young Madeira wines, since unique aroma descriptors were linked to grape variety. Malvasia and Boal grapes are associated to almond and cocoa aroma descriptors, while Madeira wine obtained from Sercial and Verdelho grapes is associated to mushroom and honey descriptors [52].

The singular and unique characteristics of Madeira wines result from the particularities of the winemaking process (Figure 5). A natural grape spirit is added in order to stop the fermentation process and to obtain an ethanol content between 18 and 22% [1,35]. On the basis of the fermentation time, Madeira wine covers four wine types: dry (attained from Sercial, total sugars: 49.1 to 64.8 g/L), medium dry (Verdelho, 64.8 to 80.4 g/L), medium sweet (Boal, 80.4 to 96.1 g/L), and sweet (Malvasia, 96.1 to 150 g/L). Tinta Negra is used to produce all types of Madeira wines [1].

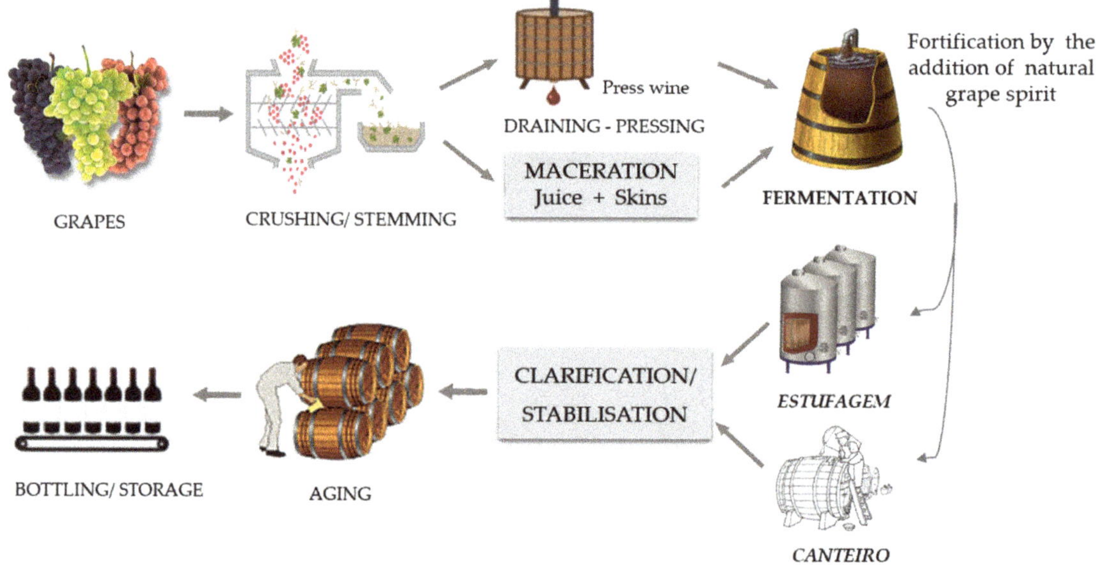

Figure 5. Schematic diagram of the winemaking process of Madeira wine.

During the fermentation beyond ethanol, glycerol, and diols, several other fermentative-derived VOCs are produced by yeast metabolism, alcohols, esters, and acids being the most dominant. The predominant alcohols detected in Madeira wines are 3-Methylbutan-1-ol, 2-phenylethanol, 2-ethylhexan-1-ol, benzyl alcohol, and butan-2,3-diol [52–54]. Both 3-Methylbutan-1-ol and 2-phenylethanol have been reported to contribute positively to Madeira wine aroma with fruit, flower, and honey descriptors [4,52,54,55]. Furthermore, 2-ethylhexan-1-ol can explain the citrus descriptors of Madeira wines obtained for Malvasia, Sercial and Tina Negra grape varieties [52].

The ethyl esters of short- and medium-chain-length carboxylic acid (C_2-C_{10}) and acetates of short-chain-length alcohols (C_4-C_6) have been detected in Madeira wines at concentration higher than their OTs (few µg/L) [1]. These fermentative-derived VOCs contribute positively to Madeira wine aroma with fruity and/or floral aroma descriptors. Ethyl acetate, isoamyl acetate, ethyl hexanoate, and 2-phenyethyl acetate were the most significant esters for overall Madeira wine aroma. Additionally, the citrus descriptor characteristic of Malvasia, Sercial, and Tinta Negra young Madeira wines could be associated to the presence of hexyl acetate and ethyl 3-methylbutanoate [54,56].

Acetic acid is the predominant volatile acid in Madeira wines, indicating around 90% of total volatile acid fraction. When present at concentrations higher than 0.7 g/L, it contributes negatively with a vinegar-like descriptor to the wine aroma. Hexanoic, octanoic, and decanoic acids are the most predominant fatty acids found in Madeira wine [53,54]. The occurrence of these acids in wines at high concentrations is associated to negative aroma descriptors such as rancid, cheesy, and vinegar. In Madeira wines they normally occur at concentrations below their OTs [1].

2.4. *Marsala Wine*

This Italian wine is produced exclusively in the Marsala PDO region (the western part of Sicily Island). Marsala (15–20% alcohol by volume) is produced using Catarratto, Damaschino, and Grillo cultivars (white grapes), but red ones can be equally used to obtain ruby-colored wines (e.g., Pignatello, Perricone, or Calabrese) [37,57,58]. Although they are usually vinified as still wine, Marsalas are subject to significant pressing, thus obtaining higher amount of dry extracts (25–30 g/L) and oxidizable substances [58]. The fermentation is usually performed at a controlled temperature between 18–20 °C. According to the desired level of sweetness, the fortification step can be accomplished either during or after fermentation [57]. Fortification can be performed by using neutral grape spirit, brandy or mistelle (adding alcohol to unfermented grape juice or partially fermented wine). Marsala wines often maturate in wooden barrels in a similar way to the solera system. Marsalas are classified in dry, medium-dry, and sweet in relation to the degree of sweetness.

2.5. *Moscatel de Setúbal*

This fortified wine, of about 17% alcohol by volume, is produced in the Setúbal PDO region, Portugal. The two traditional designations, "Moscatel Roxo" and "Moscatel de Setúbal", are produced using at least 85% of the corresponding cultivars. In addition, there are other varieties which can be used for blending (e.g., Diagalves, Boais, or Arinto) [59]. A short fermentation period is characteristic of their vinification, which is subsequently interrupted (grape spirit fortification), followed by a period of maceration (prolonged contact between skins and fortified wine). According to the grapes used, Moscatel de Setúbal wines can be classified as white or red styles. Considering the aging time, wines are categorized as young or classic (up to 5 and more than 5 years, respectively).

2.6. *Vermouth Wine*

Vermouth is a particular type of fortified wine (15–21% alcohol by volume). Its main feature is the fact that this is also flavored by using spices, herbs, or their extract to a base wine [40,60]. Among the flavoring agents, wormwood, cloves, coriander, and chamomile are most commonly used. The fortification step begins after blending the base wine (preferentially neutral and produced from white grapes) with the extracts, and before aging. The period of maturation lasts an average of 4–5 years, although more aged Vermouth wines exist. Traditionally, Italian and French styles Vermouths are the most popular worldwide, however other types can also be found. The Italian Vermouths are sweet, showing an alcoholic content ranging between 15–17% alcohol by volume; conversely, the French ones are dry, reaching approximately 18% alcohol by volume. Generally, due to the contribution of the flavoring agents, Vermouths are characterized by a bitter aftertaste and a pleasant, intense flavor.

2.7. *Commandaria Wine*

This is the sweet wine par excellence of the island of Cyprus, produced in the Commandaria PDO region [61]. It is made by using Mavro and Négrette (sun-dried red) and Xynisteri (white) grape varieties. Due to the high levels of alcohol produced (approximately

15%), must fermentation is naturally ceased. Even if the fortification is not mandatory, the resulting wine can be fortified up to 20% alcohol by volume. Aging period is performed for at least 4 years in oak barrels following a solera-like system. These dark wines are characterized by honey and raisin notes.

3. Volatile Composition of Fortified Wines

3.1. Sherry and Port Wines

In a previous study, the key role of acetaldehyde in Sherry wines, proving to be a fundamental compound for their authenticity, has been reported [62]. Generally, acetaldehyde is abundant in fortified wines, and it is responsible for the pungent feature of *Finos* and for the ripe apple notes of Sherries. Acetaldehyde is also a precursor of other key odorants, such as 2,3-butanediol and acetoin (buttery notes), 1,1-diethoxyethane (green and fruity notes) and sotolon (nutty notes). In addition to these features, this compound is also responsible for the browning phenomenon that occurs in Sherry wines. Among the VOCs characteristic of *Fino* Sherry-type wines (biological aging) it is possible to find fruity notes due to ethyl octanoate (banana, pineapple), sotolon, acetaldehyde, 1,1-diethoxyethane, and ethyl acetate (pineapple). There are also spicy notes due to the compounds released by wood, (Z)-oak lactone (vanilla), eugenol and 4-ethylguaiacol (clove), and sotolon [63].

Other studies have highlighted the characteristic compounds of *Oloroso* wines (oxidative aging). Chemical, vegetable, fruity, balsamic, floral, and spicy tones represent the eight odorant series established for their description [64,65]. Considering *Oloroso* wines age between 0 and 14 years, ethyl butanoate (banana), sotolon, and ethyl octanoate were found to be the most active odorants. Acetaldehyde, 2,3-butanedione, and 1,1-diethoxyethane increased their impact during aging, rather than compounds from cask wood. If compared with *Fino*- and *Oloroso*-type wines, Amontillado showed that acetaldehyde, eugenol, and ethyl acetate were the compounds that most distinguish these three types of Sherry wines [66,67]. Significant changes in the aromatic profile of these wines were reported during the first years of the oxidative period. For Amontillado wines, it was concluded that the most representative odorant compounds are ethyl octanoate, ethyl butanoate, sotolon, eugenol, and ethyl isobutyrate. Another study, focused on the characterization of some hydroxy acids in different types of wines, including Sherries, showed that the latter had a concentration of these compounds between 2 and 40 times higher compared to all other samples [68]. The differences in concentration with respect to other types of wine are particularly significant in the case of 2-hydroxy-3-methylbutanoic acid and 2-hydroxy-2-methylbutanoic acid. This evidence might be related to the typical vinification and aging processes of Sherry wines.

Considering Port wines, some studies focused on specific compounds formed during the Maillard reaction, particularly aldehydes and VOCs formed from carotenoid degradation. Another line of research has focused on the volatile compounds that form or vary during the aging process. Regarding the first group, for example, it has proven the strong link between β-carotene, lutein, and related compounds (β-damascenone, β-cyclocitral, and β-ionone), responsible for the sensorial impact of Port wines [69]. Interestingly, the branched aldehydes (dried fruit notes) and (*E*)-2-alkenals showed significant amounts in Ports (both Tawny and Ruby wines) [70]. In another study, the 3-deoxyosone content was evaluated during both the natural and forced aging samples. Interestingly, it emerged that this compound can be used as a wine aging marker due to the relationship between its yield and the reduction in sugars and amino acids in these wines during the Maillard reaction [71]. Due to the presence of high alcohol levels, Port wines showed high levels of VOCs extracted from wood (oak aging), in particular of oak lactone (β–methyl-γ-octalactone) [47]. It has been reported that one of the most significant aromatic compounds is sotolon. Both temperature and oxygen levels can influence the formation of sotolon (nutty and spicy notes) [72]. Older Ports showed less volatile sulphur compounds, such as 3-(methylthio)-1-propanol (cauliflower), 2-(methylthio)ethanol (French bean), and 3-(methylthio)-1-propionic acid (butter) when compared with young wines. Due to exposure to oxygen, aged Port wines show higher amounts of dimethyl sulphide, responsible for quince and truffle notes. In Tawnies, the aging process also influences the content of nor-isoprenoids, namely increasing vitispirane, 1,1,6-trimethyl-1,2-dihydronaphthalene (TDN) and 2,2,6-trimethylcyclohexanone (TCH). Similarly, the concentrations of β-ionone and β-damascenone are affected in vintage wines, and this behavior is related to oxygen exposure in bottle and cask aging [73].

3.2. Madeira Wine

Highest quality Madeira wines are typically aged in *canteiro* (the wine barrels are placed under supports of wooden beams, called beds), which involve aging the wine in oak casks for least 3 years at temperatures ranging from 15 to 31 °C, in high humidity conditions (>70%), before being marketed. Nevertheless, a significant percentage of Madeira wines are submitted to the *estufagem* process, in which the wine is initially submitted to a thermal process for at least 3 months, during which the temperature gradually rises 5 °C/day and kept at 45–50 °C. After this period, the wines are submitted to a maturation step in oak casks for a least of 3 years before being marketed [74,75]. Madeira wine's specific sensorial properties are strongly dependent on these production techniques, which are responsible for numerous reactions, mainly the Maillard reaction, Strecker degradation, and caramelization, influenced by a combination of numerous factors including pH, temperature, time, and dissolved oxygen and SO_2, in addition to the diffusion of compounds (volatile substances and ellagitannins from the oak to the wine. The nature of the diffused compounds depends primarily on the oak properties (e.g., geographic origin, oak species, toasting degree, and oak age) [35]. Both enzymatic and non-enzymatic reactions occurring during winemaking are responsible for the main differences in the sensorial perception throughout the process. Pereira et al. [53] evaluated the behavior of the volatile pattern of Madeira wines submitted to the *estufagem* process (45 °C, 3 months) and to an overheating process (70 °C, 1 month). The obtained results indicated that accelerated aging favored the development of volatiles, such as phenylacetaldehyde, β-damascenone, and 5-(ethoxymethyl)-2-furfural. In addition, numerous varietal compounds (e.g., monoterpenic alcohols) responsible for the floral descriptors of some Madeira wines were not found after thermal treatment. Miranda et al. [75] evaluated the trend of acetic acid and ethyl acetate during the aging process of Madeira wine using both *estufagem* (forced aging associated with thermal treatment) and *canteiro* (aging in oak casks). The data indicated that these two VOCs showed a similar formation trend in both types of aging processes. More recently, Perestrelo et al. [54] evaluated the influence of sugar content, storage time (0, 1, and 4 months), and temperature (30, 45, and 55 °C) on volatile profile of Tinta Negra wines. A total of 65 VOCs were identified, from which 14 appear during storage. This result suggested that storage promotes the overall aroma of Tinta Negra wines. Additionally, during the aging process the wines acquired aroma descriptors (e.g., caramel, dried fruit, spice, toast, and wood) as a result of the Maillard reaction, Strecker degradation, caramelization, and microbial activity, whereas lost their fresh, floral, and fruity descriptors characteristic of younger wines (Figure 6). The results obtained in this research represent a useful tool to promote alterations in the *estufagem* process, as well as assess the effect of storage on volatile profile.

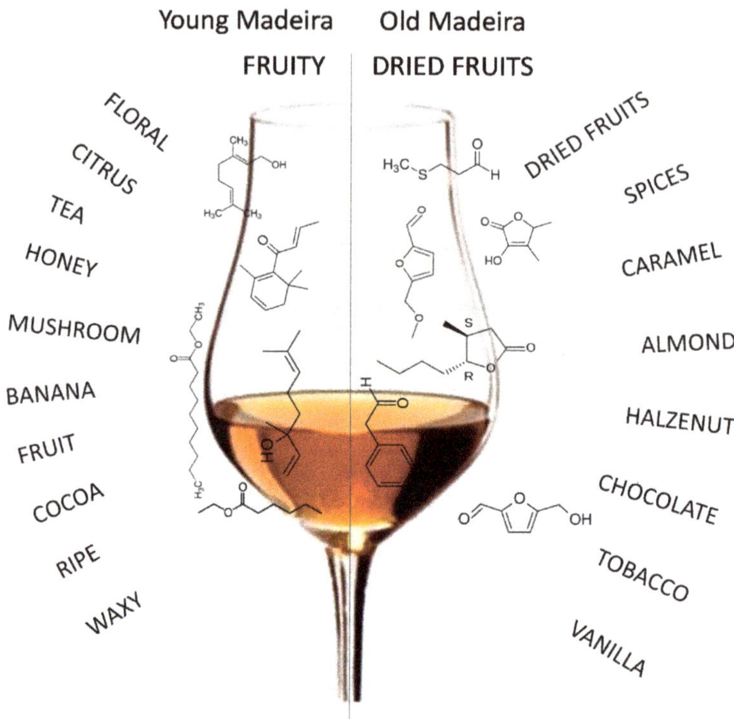

Figure 6. Characteristic aroma notes in young and old Madeira wines.

Wine volatile pattern represents a significant role in wine quality, as its volatiles endorse numerous sensations through wine drink, odors, and tastes that lead to consumer acceptance and/or rejection. Overall, an occurrence of a single molecule, at a concentration higher than its OT, is sufficient to deliver a distinctive aroma (key odorant). Nonetheless, the key odorant comprises of a mixture of hundreds of diverse odorants (VOCs). Numerous VOCs are present in only trace levels (frequently from µg/L or ng/L level), but, even at concentrations lower their OT, they may donate to the global aroma; due to their combinations with other VOCs, they can consequently have an influence on wine aroma perception [1,52,76]. Moreover, several investigations have centered on the identification of putative key odorants that can provide independently to the overall Madeira wine aroma [4,77]. Campo et al. [4] established the aroma pattern of four Madeira wines from the grape varieties Malvasia, Boal, Verdelho, and Sercial. The gas chromatography-olfactometry (GC–O) profile of Madeira wines lack the most significant varietal aromas, such as linalool, 3-mercaptohexyl acetate, and methoxypyrazines, whereas are rich in sotolon, phenylacetaldehyde, and wood extractable aromas. They also contain a great number of powerful odorants not recognized, which were not even found in the matching young wines. Moreover, furanic compounds, such as 2-furfural, 5-methyl-2-furfural, 5-hydroxymethyl-2-furfural, and 5-(ethoxymethyl)-2-furfural are quantitatively significant in Madeira wines analyzed, but they were not identified by GC–O, which indicate that furanic compounds probably do not contribute significantly to Madeira wine aroma due to their high OT. In addition, the highest scores detected were related to candy, toasty, woody, and dried fruit descriptors. Nevertheless, these depend on grape variety, winemaking, and the aging process. Silva et al. [77] evaluated the effect of forced-aging on Madeira wine aroma by GC-O. Several Maillard by-products, such as sotolon, 2-furfural, 5-methyl-2-furfural, 5-(ethoxymethyl)-2-furfural, methional, and phenylacetaldehyde were detected, which clarify the baked, brown sugar, and nutty aroma descriptors of Madeira wine. From these studies [4,77], sotolon, due to its low OT (10 µg/L), is reported as a key odorant on the conventional aroma of aged Madeira wines [36,78,79]. The formation pathways of sotolon are not well elucidated [79], although many precursors and pathways have been suggested for its formation in wines, namely oxidative degradation of ascorbic acid in the presence of ethanol, aldol condensation, peroxidation of acetaldehyde, and the Maillard reaction, among others [36,78]. Nevertheless, it is documented that the sugar content, wine oxidation, storage time, and temperature are strongly linked to sotolon formation [78,80]. According to Câmara et al. [36], Malvasia has a higher level of sotolon when compared with Sercial, and the study observed that sotolon amount is dependent of aging, sugar content, and furanic compounds. Some lactones derived from oak, such as γ-butyrolactone, pantolactone, *cis* and *trans*-whisky lactone, represent important odorants in Madeira wines aged in oak casks. During the aging process in oak casks, their concentration tends to increase notably [20].

4. Conclusions

The current review is focused on the most worldwide fortified wines—Madeira, Port, Sherry, Vermouth and Muscat—which are produced using traditional and particular vinification procedures. The distinctive odor descriptors of these fortified wines result from the integration of different factors such as grape varieties, geographical region, vinification, and aging processes. VOCs are formed in different ways and in the different stages that occur; from the formation of the grapes to when the final consumer–terpenoids, C13 norisoprenoids, and sesquiterpenoids are biosynthesized in the grapes, alcohols, esters, and acids are the result of the microorganisms actions, while lactones, furanic compounds acetals, and volatile phenols, are synthesized during aging. Nevertheless, the impact of these VOCs to the global aroma of fortified wines depends on their concentration and OTs. According to previous studies, reported in the current review, during aging the fortified wines lose their freshness and fruitiness odor descriptors (primary and secondary aromas) and other odor descriptors are formed, namely spicy, caramel, toast, wood, and dried fruits, as result of the Maillard reaction and diffusion from the oak to the wines. Sotolon, oak lactones, 2-furfural, 5-methyl-2-furfural, 5-(ethoxymethyl)-2-furfural, and 5-phenylacetaldehyde are some VOCs that could explain the odor descriptors related to wine aging.

Author Contributions: Writing—original draft preparation, T.A., J.S.C. (Section 1), M.B., M.L., J.D.C. (Section 2); R.P. (Section 3); Writing—review and editing, R.P., J.S.C.; Supervision, J.S.C. All authors have read and agreed to the published version of the manuscript.

Funding: This work was supported by FCT-Fundação para a Ciência e a Tecnologia through the CQM Base Fund—UIDB/00674/2020, and Programmatic Fund—UIDP/00674/2020, and by Madeira 14–20 Program, project PROEQUIPRAM—Reforço do Investimento em Equipamentos e Infraestruturas Científicas na RAM (M1420-01-0145-FEDER-000008), by ARDITI—Agência Regional para o Desenvolvimento da Investigação Tecnologia e Inovação, through the projects M1420-01-0145-FEDER-000005—Centro de Química da Madeira—CQM$^+$ (Madeira 14–20 program).

Conflicts of Interest: The authors declare no conflict of interest.

References

1. Perestrelo, R.; Silva, C.; Gonçalves, C.; Castillo, M.; Câmara, J.S. An Approach of the Madeira Wine Chemistry. *Beverages* **2020**, *6*, 12. [CrossRef]
2. Mina, M.; Tsaltas, D. Contribution of Yeast in Wine Aroma and Flavour. In *Yeast—Industrial Applications*; IntechOpen: London, UK, 2017.
3. Lin, J.; Massonnet, M.; Cantu, D. The genetic basis of grape and wine aroma. *Hortic. Res.* **2019**, *6*, 1–24. [CrossRef]
4. Campo, E.; Ferreira, V.; Escudero, A.; Marqués, J.C.; Cacho, J. Quantitative gas chromatography—Olfactometry and chemical quantitative study of the aroma of four Madeira wines. *Anal. Chim. Acta* **2006**, *563*, 180–187. [CrossRef]
5. Lytra, G.; Tempere, S.; Le Floch, A.; De Revel, G.; Barbe, J.-C. Study of Sensory Interactions among Red Wine Fruity Esters in a Model Solution. *J. Agric. Food Chem.* **2013**, *61*, 8504–8513. [CrossRef]

6. Ong, P.K.C.; Acree, T.E. Similarities in the Aroma Chemistry of Gewürztraminer Variety Wines and Lychee (*Litchi chinesis* Sonn.). *Fruit J. Agric. Food Chem.* **1999**, *47*, 665–670. [CrossRef]
7. Díaz-Maroto, M.C.; Guchu, E.; Castro-Vázquez, L.; de Torres, C.; Pérez-Coello, M.S. Aroma-active compounds of American, French, Hungarian and Russian oak woods, studied by GC–MS and GC–O. *Flavour Fragr. J.* **2008**, *23*, 93–98. [CrossRef]
8. Genovese, A.; Gambuti, A.; Piombino, P.; Moio, L. Sensory properties and aroma compounds of sweet Fiano wine. *Food Chem.* **2007**, *103*, 1228–1236. [CrossRef]
9. Tarasov, A.; Giuliani, N.; Dobrydnev, A.; Schuessler, C.; Volovenko, Y.; Rauhut, R.; Jung, R. 1,1,6-Trimethyl-1,2-dihydronaphthalene (TDN) Sensory Thresholds in Riesling Wine. *Foods* **2020**, *9*, 606. [CrossRef]
10. Mafata, M.; Stander, M.A.; Thomachot, B.; Buica, A. Measuring Thiols in Single Cultivar South African Red Wines Using 4,4-Dithiodipyridine (DTDP) Derivatization and Ultraperformance Convergence Chromatography-Tandem Mass Spectrometry. *Foods* **2018**, *7*, 138. [CrossRef]
11. Gaby, J.M.; Bakke, A.J.; Baker, A.N.; Hopfer, H.; Hayes, J.E. Individual Differences in Thresholds and Consumer Preferences for Rotundone Added to Red Wine. *Nutrients* **2020**, *12*, 2522. [CrossRef]
12. Perestrelo, R.; Barros, A.S.A.S.; Rocha, S.M.; Câmara, J.S. Establishment of the varietal profile of *Vitis vinifera* L. grape varieties from different geographical regions based on HS-SPME/GC-qMS combined with chemometric tools. *Microchem. J.* **2014**, *116*, 107–117. [CrossRef]
13. Moreno-Arribas, M.V.; Polo, M.C. Winemaking Biochemistry and Microbiology: Current Knowledge and Future Trends. *Crit. Rev. Food Sci. Nutr.* **2005**, *45*, 265–286. [CrossRef]
14. Lukić, I.; Horvat, I. Differentiation of Commercial PDO Wines Produced in Istria (Croatia) According to Variety and Harvest Year Based on HS-SPME-GC/MS Volatile Aroma Compound Profiling. *Food Technol. Biotechnol.* **2017**, *55*, 95–108. [CrossRef]
15. Huang, X.-Y.; Jiang, Z.-T.; Tan, J.; Li, R. Geographical Origin Traceability of Red Wines Based on Chemometric Classification via Organic Acid Profiles. *J. Food Qual.* **2017**, *2017*, 2038073. [CrossRef]
16. Perestrelo, R.; Silva, C.; Câmara, J.S. A useful approach for the differentiation of wines according to geographical origin based on global volatile patterns. *J. Sep. Sci.* **2014**, *37*, 1974–1981. [CrossRef] [PubMed]
17. Oliveira, J.M.; Faria, M.; Sá, F.; Barros, F.; Araújo, I.M. C6-alcohols as varietal markers for assessment of wine origin. *Anal. Chim. Acta* **2006**, *563*, 300–309. [CrossRef]
18. González-Barreiro, C.; Rial-Otero, R.; Cancho-Grande, B.; Simal-Gándara, J. Wine aroma compounds in grapes: A critical review. *Crit. Rev. Food Sci. Nutr.* **2015**, *55*, 202–218. [CrossRef]
19. Belda, I.; Ruiz, J.; Esteban-Fernández, A.; Navascués, E.; Marquina, D.; Santos, A.; Moreno-Arribas, M. Microbial contribution to wine aroma and its intended use for wine quality improvement. *Molecules* **2017**, *22*, 189. [CrossRef]
20. Câmara, J.; Alves, A.; Marques, J. Changes in volatile composition of Madeira wines during their oxidative ageing. *Anal. Chim. Acta* **2006**, *563*, 188–197. [CrossRef]
21. Le Menn, N.; Van Leeuwen, C.; Picard, M.; Riquier, L.; De Revel, G.; Marchand, S. Effect of Vine Water and Nitrogen Status, as Well as Temperature, on Some Aroma Compounds of Aged Red Bordeaux Wines. *J. Agric. Food Chem.* **2019**, *67*, 7098–7109. [CrossRef] [PubMed]
22. Zhu, F.; Du, B.; Li, J. Aroma Compounds in Wine. In *Grape and Wine Biotechnology*; Morata, A., Ed.; IntechOpen: London, UK, 2016; ISBN 9789535126935.
23. Metafa, M.; Economou, A. Comparison of solid-phase extraction sorbents for the fractionation and determination of important free and glycosidically-bound varietal aroma compounds in wines by gas chromatography-mass spectrometry. *Open Chem.* **2013**, *11*, 228–247. [CrossRef]
24. Ribéreau-Gayon, P.; Glories, Y.; Maujean, A.; Dubourdieu, D. *Handbook of Enology, The Chemistry of Wine: Stabilization and Treatments*, 2nd ed.; John Wiley & Sons: Chichester, UK, 2006; Volume 2, ISBN 9780470010396.
25. Sidhu, D.; Lund, J.; Kotseridis, Y.; Saucier, C. Methoxypyrazine Analysis and Influence of Viticultural and Enological Procedures on their Levels in Grapes, Musts, and Wines. *Crit. Rev. Food Sci. Nutr.* **2014**, *55*, 485–502. [CrossRef] [PubMed]
26. Padilla, B.; Gil, J.V.; Manzanares, P. Past and Future of Non-Saccharomyces Yeasts: From Spoilage Microorganisms to Biotechnological Tools for Improving Wine Aroma Complexity. *Front. Microbiol.* **2016**, *7*, 411. [CrossRef]
27. Mozzon, M.; Savini, S.; Boselli, E.; Thorngate, J.H. The herbaceous character of wines. *Ital. J. Food Sci.* **2016**, *28*, 190–207.
28. Divol, B.; Strehaiano, P.; Lonvaud-Funel, A. Effectiveness of dimethyldicarbonate to stop alcoholic fermentation in wine. *Food Microbiol.* **2005**, *22*, 169–178. [CrossRef]
29. Lasik-Kurdyś, M.; Majcher, M.; Nowak, J. Effects of Different Techniques of Malolactic Fermentation Induction on Diacetyl Metabolism and Biosynthesis of Selected Aromatic Esters in Cool-Climate Grape Wines. *Molecules* **2018**, *23*, 2549. [CrossRef] [PubMed]
30. Chisholm, M.G.; Guiher, L.A.; Zaczkiewicz, S.M. Aroma characteristics of aged Vidal blanc wine. *Am. J. Enol. Vitic.* **1995**, *46*, 56–62.
31. Koussissi, E.; Dourtoglou, V.; Ageloussis, G.; Paraskevopoulos, Y.; Dourtoglou, T.; Paterson, A.; Chatzilazarou, A. Influence of toasting of oak chips on red wine maturation from sensory and gas chromatographic headspace analysis. *Food Chem.* **2009**, *114*, 1503–1509. [CrossRef]

32. Perestrelo, R.; Barros, A.S.; Camara, J.S.; Rocha, S.M. In-Depth Search Focused on Furans, Lactones, Volatile Phenols, and Acetals as Potential Age Markers of Madeira Wines by Comprehensive Two-Dimensional Gas Chromatography with Time-of-Flight Mass Spectrometry Combined with Solid Phase Microextraction. *J. Agric. Food Chem.* **2011**, *59*, 3186–3204. [CrossRef]
33. Echave, J.; Barral, M.; Fraga-Corral, M.; Prieto, M.; Simal-Gandara, J. Bottle Aging and Storage of Wines: A Review. *Molecules* **2021**, *26*, 713. [CrossRef]
34. Pereira, A.C.; Reis, M.S.; Saraiva, P.M.; Marqués, J.C. Madeira wine ageing prediction based on different analytical techniques: UV–vis, GC-MS, HPLC-DAD. *Chemom. Intell. Lab. Syst.* **2011**, *105*, 43–55. [CrossRef]
35. Perestrelo, R.; Nogueira, J.M.F.; Câmara, J.S. Potentialities of two solventless extraction approaches—Stir bar sorptive extraction and headspace solid-phase microextraction for determination of higher alcohol acetates, isoamyl esters and ethyl esters in wines. *Talanta* **2009**, *80*, 622–630. [CrossRef] [PubMed]
36. Câmara, J.S.; Marques, J.C.; Alves, A.; Ferreira, A.C.S. 3-Hydroxy-4,5-dimethyl-2(5H)-furanone Levels in Fortified Madeira Wines: Relationship to Sugar Content. *J. Agric. Food Chem.* **2004**, *52*, 6765–6769. [CrossRef]
37. Tredoux, A.; Ferreira, A.C.S. Fortified Wines: Styles, Production and Flavour Chemistry. In *Alcoholic Beverages*; Elsevier: Amsterdam, The Netherlands, 2012; pp. 159–179.
38. Reader, H.P.; Dominguez, M. Fortified Wines Sherry, Port and Madeira. In *Fermented Beverage Production*; Springer: Berlin/Heidelberg, Germany, 2003; pp. 157–194.
39. Pozo-Bayón, M.Á.; Moreno-Arribas, M.V. Sherry Wines: Manufacture, Composition and Analysis. In *Encyclopedia of Food and Health*; Elsevier: Amsterdam, The Netherlands, 2016; pp. 779–784.
40. Joshi, V.K.; Sharma, S.; Thakur, A.D. Wines: White, Red, Sparkling, Fortified, and Cider. In *Current Developments in Biotechnology and Bioengineering: Food and Beverages Industry*; Elsevier: Amsterdam, The Netherlands, 2016; pp. 353–406. ISBN 9780444636775.
41. Jackson, R. Innovations in Winemaking. In *Science and Technology of Fruit Wine Production*; Elsevier: Amsterdam, The Netherlands, 2017; pp. 617–662. ISBN 9780128010341.
42. Jackson, R. Specific and Distinctive Wine Styles. In *Wine Science*; Elsevier: Amsterdam, The Netherlands, 2014; pp. 677–759.
43. Bamforth, C.W.; Cook, D.J. Fortified Wines. In *Food, Fermentation, and Micro-Organisms*; John Wiley & Sons: Chichester, UK, 2019; pp. 111–115.
44. Zea, L.; Moyano, L.; Moreno, J.; Cortes, B.; Medina, M. Discrimination of the aroma fraction of Sherry wines obtained by oxidative and biological ageing. *Food Chem.* **2001**, *75*, 79–84. [CrossRef]
45. Pozo-Bayón, M.A.; Moreno-Arribas, M.V. Sherry Wines. In *Advances in Food and Nutrition Research*; Academic Press: Cambridge, MA, USA, 2011; Volume 63, pp. 17–40.
46. Jackson, R. Post-Fermentation Treatments and Related Topics. In *Wine Science*; Elsevier: Amsterdam, The Netherlands, 2014; pp. 535–676.
47. Bakker, J.; Clarke, R.J. Sherry, Port and Madeira. In *Wine Flavour Chemistry*; Wiley-Blackwell: Oxford, UK, 2011; pp. 291–339.
48. Moreira, N.; Guedes de Pinho, P. Port Wine. In *Advances in Food and Nutrition Research*; Academic Press: Cambridge, MA, USA, 2011; Volume 63, pp. 119–146.
49. Hogg, T. Port. In *Sweet, Reinforced and Fortified Wines*; John Wiley Sons: Oxford, UK, 2013; pp. 305–317.
50. Perestrelo, R.; Albuquerque, F.; Rocha, S.; Câmara, J. Distinctive Characteristics of Madeira Wine Regarding Its Traditional Winemaking and Modern Analytical Methodologies. In *Advances in Food and Nutrition Research*; Elsevier: Amsterdam, The Netherlands, 2011; Volume 63, pp. 207–249.
51. Perestrelo, R.; Barros, A.S.; Rocha, S.; Câmara, J. Optimisation of solid-phase microextraction combined with gas chromatography–mass spectrometry based methodology to establish the global volatile signature in pulp and skin of *Vitis vinifera* L. grape varieties. *Talanta* **2011**, *85*, 1483–1493. [CrossRef] [PubMed]
52. Perestrelo, R.; Silva, C.; Câmara, J.S. The sui generis and notable peculiarities of fortified wines: common characteristics and major differences. In *Encyclopedia of Food and Health*; Caballero,B., Finglas, P.M., Toldrá, F., Eds.; Academic Press: Amsterdam, The Netherlands, 2016; Chapter 758 Wines | Fortified Wines. 534-555 ISBN: 978-0-12-3839472.
53. Pereira, V.; Cacho, J.; Marques, J.C. Volatile profile of Madeira wines submitted to traditional accelerated ageing. *Food Chem.* **2014**, *162*, 122–134. [CrossRef]
54. Perestrelo, R.; Silva, C.L.; Silva, P.; Câmara, J.S. Impact of storage time and temperature on volatomic signature of Tinta Negra wines by LLME/GC-ITMS. *Food Res. Int.* **2018**, *109*, 99–111. [CrossRef]
55. Perestrelo, R.; Fernandes, A.; Albuquerque, F.; Marques, J.; Câmara, J. Analytical characterization of the aroma of Tinta Negra Mole red wine: Identification of the main odorants compounds. *Anal. Chim. Acta* **2006**, *563*, 154–164. [CrossRef]
56. Perestrelo, R.; Silva, C.; Câmara, J.S. Madeira wine volatile profile. A platform to establish madeira wine aroma descriptors. *Molecules* **2019**, *24*, 3028. [CrossRef] [PubMed]
57. Ferreira, I.M.; Pérez-Palacios, M.T. Anthocyanic Compounds and Antioxidant Capacity in Fortified Wines. In *Processing and Impact on Antioxidants in Beverages*; Elsevier: Amsterdam, The Netherlands, 2014; pp. 3–14. ISBN 9780124047389.
58. Zanfi, A.; Mencarelli, S. Marsala. In *Sweet, Reinforced and Fortified Wines*; John Wiley & Sons: Oxford, UK, 2013; pp. 319–325.
59. Feliciano, R.P.; Bravo, M.N.; Pires, M.M.; Serra, A.T.; Duarte, C.M.; Boas, L.V.; Bronze, M.R. Phenolic Content and Antioxidant Activity of Moscatel Dessert Wines from the Setúbal Region in Portugal. *Food Anal. Methods* **2009**, *2*, 149–161. [CrossRef]
60. Panesar, P.S.; Joshi, V.K.; Panesar, R.; Abrol, G.S. Vermouth: Technology of production and quality characteristics. In *Advances in Food and Nutrition Research*; Academic Press: Cambridge, MA, USA, 2011; Volume 63, pp. 251–283.

61. Mencarelli, F. Notes on Other Sweet Wines. In *Sweet, Reinforced and Fortified Wines: Grape Biochemistry, Technology and Vinification*; John Wiley & Sons: Hoboken, NJ, USA, 2013; pp. 327–329. ISBN 9780470672242.
62. Zea, L.; Serratosa, M.P.; Merida, J.; Moyano, L. Acetaldehyde as Key Compound for the Authenticity of Sherry Wines: A Study Covering 5 Decades. *Compr. Rev. Food Sci. Food Saf.* **2015**, *14*, 681–693. [CrossRef]
63. Zea, L.; Moyano, L.; Moreno, J.A.; Medina, M. Aroma series as fingerprints for biological ageing in fino sherry-type wines. *J. Sci. Food Agric.* **2007**, *87*, 2319–2326. [CrossRef]
64. Zea, L.; Moyano, L.; Medina, M. Changes in aroma profile of sherry wines during the oxidative ageing. *Int. J. Food Sci. Technol.* **2010**, *45*, 2425–2432. [CrossRef]
65. Zea, L.; Moyano, L.; Ruiz, M.J.; Medina, M. Odor Descriptors and Aromatic Series During the Oxidative Aging of Oloroso Sherry Wines. *Int. J. Food Prop.* **2013**, *16*, 1534–1542. [CrossRef]
66. Marcq, P.; Schieberle, P. Characterization of the Key Aroma Compounds in a Commercial Amontillado Sherry Wine by Means of the Sensomics Approach. *J. Agric. Food Chem.* **2015**, *63*, 4761–4770. [CrossRef] [PubMed]
67. Ruiz-Bejarano, M.J.; Castro-Mejías, R.; Rodríguez-Dodero, M.D.C.; García-Barroso, C. Volatile composition of Pedro Ximénez and Muscat sweet Sherry wines from sun and chamber dried grapes: A feasible alternative to the traditional sun-drying. *J. Food Sci. Technol.* **2016**, *53*, 2519–2531. [CrossRef] [PubMed]
68. Gracia-Moreno, E.; Lopez, R.; Ferreira, V. Quantitative determination of five hydroxy acids, precursors of relevant wine aroma compounds in wine and other alcoholic beverages. *Anal. Bioanal. Chem.* **2015**, *407*, 7925–7934. [CrossRef]
69. Ferreira, A.C.S.; Monteiro, J.; Oliveira, C.; De Pinho, P.G.; Pedroso, J.M. Study of major aromatic compounds in port wines from carotenoid degradation. *Food Chem.* **2008**, *110*, 83–87. [CrossRef]
70. Culleré, L.; Cacho, A.J.; Ferreira, V. An Assessment of the Role Played by Some Oxidation-Related Aldehydes in Wine Aroma. *J. Agric. Food Chem.* **2007**, *55*, 876–881. [CrossRef]
71. Oliveira, C.M.; Santos, S.A.; Silvestre, A.J.; Barros, A.S.; Ferreira, A.C.S.; Silva, A.M. Quantification of 3-deoxyglucosone (3DG) as an aging marker in natural and forced aged wines. *J. Food Compos. Anal.* **2016**, *50*, 70–76. [CrossRef]
72. Martins, R.C.; Monforte, A.R.; Ferreira, A.S. Port Wine Oxidation Management: A Multiparametric Kinetic Approach. *J. Agric. Food Chem.* **2013**, *61*, 5371–5379. [CrossRef]
73. Prata-Sena, M.; Castro-Carvalho, B.M.; Nunes, S.; Amaral, B.; Silva, P. The terroir of Port wine: Two hundred and sixty years of history. *Food Chem.* **2018**, *257*, 388–398. [CrossRef] [PubMed]
74. Pereira, V.; Albuquerque, F.; Cacho, J.; Marques, J.C. Polyphenols, Antioxidant Potential and Color of Fortified Wines during Accelerated Ageing: The Madeira Wine Case Study. *Molecules* **2013**, *18*, 2997–3017. [CrossRef]
75. Miranda, A.; Pereira, V.; Pontes, M.; Albuquerque, F.; Marques, J.C. Acetic acid and ethyl acetate in Madeira wines: Evolution with ageing and assessment of the odour rejection threshold. *Ciência Técnica Vitivinícola* **2017**, *32*, 1–11. [CrossRef]
76. López, R.; Aznar, M.; Cacho, J.; Ferreira, V. Determination of minor and trace volatile compounds in wine by solid-phase extraction and gas chromatography with mass spectrometric detection. *J. Chromatogr. A* **2002**, *966*, 167–177. [CrossRef]
77. Silva, H.O.E.; De Pinho, P.G.; Machado, B.P.; Hogg, T.; Marques, J.C.; Câmara, J.S.; Albuquerque, F.; Ferreira, A.C.S. Impact of Forced-Aging Process on Madeira Wine Flavor. *J. Agric. Food Chem.* **2008**, *56*, 11989–11996. [CrossRef]
78. Pereira, V.; Leça, J.M.; Gaspar, J.M.; Pereira, A.C.; Marques, J.C. Rapid Determination of Sotolon in Fortified Wines Using a Miniaturized Liquid-Liquid Extraction Followed by LC-MS/MS Analysis. *J. Anal. Methods Chem.* **2018**, *2018*, 4393040. [CrossRef] [PubMed]
79. Gaspar, J.M.; Pereira, V.; Marques, A.J.C. Odor detection threshold (ODT) and odor rejection threshold (ORT) determination of sotolon in Madeira wine: A preliminary study. *AIMS Agric. Food* **2018**, *3*, 172–180. [CrossRef]
80. Pereira, V.; Santos, M.; Cacho, J.; Marques, J.C. Assessment of the development of browning, antioxidant activity and volatile organic compounds in thermally processed sugar model wines. *LWT* **2017**, *75*, 719–726. [CrossRef]

MDPI
St. Alban-Anlage 66
4052 Basel
Switzerland
Tel. +41 61 683 77 34
Fax +41 61 302 89 18
www.mdpi.com

Foods Editorial Office
E-mail: foods@mdpi.com
www.mdpi.com/journal/foods

www.ingramcontent.com/pod-product-compliance
Lightning Source LLC
LaVergne TN
LVHW070649170726
843515LV00004B/359

* 9 7 8 3 0 3 6 5 2 9 7 2 1 *